An American River

Mary Bruno

An American River

From Paradise to Superfund, Afloat on New Jersey's Passaic

Printed in the United States of America

First Printing, 2012

ISBN 978-0-615-60179-3

DeWitt Press
P.O. Box 2872, Vashon, WA 98070
www.anamericanriver.com

For the river,
And my family

Contents

The Passaic River

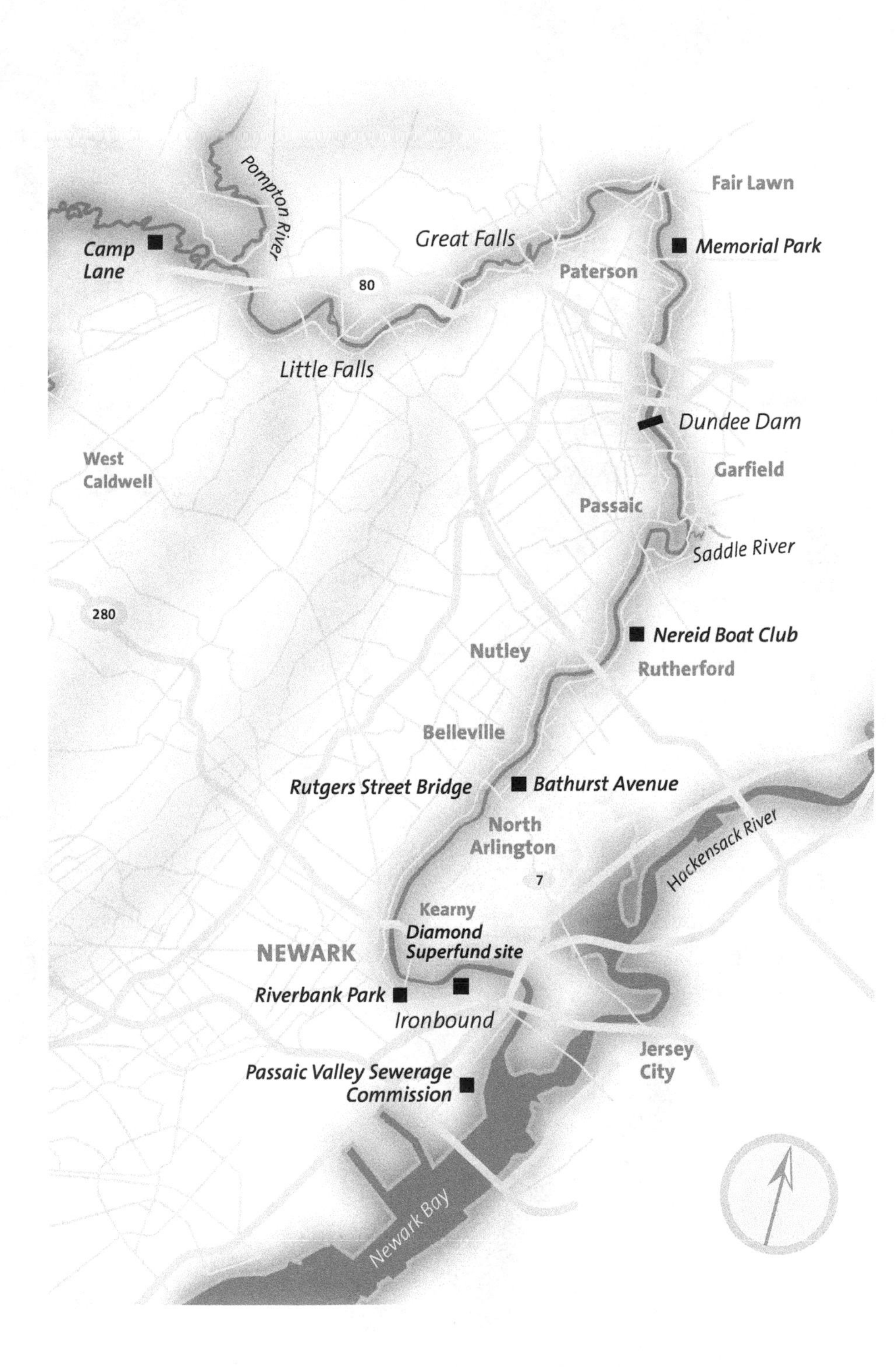

Pompton River
Fair Lawn
Camp Lane
Great Falls
Memorial Park
Paterson
80
Little Falls
Dundee Dam
West Caldwell
Garfield
Passaic
Saddle River
280
Nereid Boat Club
Nutley
Rutherford
Belleville
Rutgers Street Bridge
Bathurst Avenue
North Arlington
Hackensack River
7
Kearny
Diamond Superfund site
NEWARK
Riverbank Park
Ironbound
Jersey City
Passaic Valley Sewerage Commission
Newark Bay

An American River

Preface

For the first 18 years of my life I lived along the final 17-mile stretch of the Passaic River. That's the dirty, ugly part of the river that passes through the most crowded, industrialized part of the country. The Passaic forms the western border of my hometown, North Arlington, New Jersey, a tiny borough just a few miles north of the river's mouth in Newark. Our house sat on a steep slope above the river. In the wintertime, when the oak and maple trees were bare, we could see the water from our front porch. Sometimes in summer, when a flood tide would overwhelm the river's sluggish current, the Passaic would smell, faintly, of the sea.

I didn't have much to do with the Passaic as a kid. I crossed over it often enough, every time we visited my mother's large family, who all lived on the other side. But I never fished the Passaic or took a boat out on it. I certainly didn't swim in it. I didn't really know the river. I just knew that it gave me the creeps.

Despite my estrangement from the Passaic, and without any conscious intent, I have sought out rivers in my life. To study them, travel them, live near them. I'm drawn to the patterns and power and persistence of water on the move. In fact, I keep a river list. Only those rivers that I've been in or on qualify for my list, which is surprisingly long.

It starts with the historic Delaware, which separates New Jersey from Pennsylvania. Many summers ago, during a family picnic at my cousin Betsy's in Frenchtown, New Jersey, we went tubing down the slow, wide run of river that flows past Betsy's house.

There is the Cedar River, a gentle blackwater creek that winds through the New Jersey Pine Barrens. My brother Joe and I canoed a narrow sandy stretch of the Cedar one hot summer afternoon with George, a tenderfoot colleague of mine from *Newsweek*.

There is Ohio's Maumee River, where on a frigid winter morning I waded in to collect diatom samples. The surface waves that slapped my rubber boots turned instantly to ice, which crackled as I sloshed around in the shallows.

There's the lazy Skagit in Washington State, site of my first paddle with Kate; and the rushing Nisqually, also in Washington, where Kate and I once had a kayaking misadventure. (I insisted we could paddle downstream to the river's delta, then back upstream with the incoming tide. At least I got the first half right.)

There is the wide, muddy Missouri, the setting for a four-day-long canoe trip along the stretch that drains Montana's parched Missouri Breaks.

I have traveled up and back down Brazil's Rio Negro River, a major tributary of the Amazon; along the oh, so civilized Seine in Paris; on Africa's wild Zambezi above Victoria Falls; and through the lush inland delta that the Okavanga River creates where it empties into the Kalahari Desert. All four are on my list.

I included Washington's Columbia River, whose dramatic canyon I once water-skied through at sunset when the cliffs and sky and water were all shades of indigo; and Idaho's Snake River, where I camped with fellow birders on a hawk spotting trip; and the Rio Grande, where I cooled off once after a hike on the Texas side; and the Potomac, because on a cool

September morning many years ago I followed Ralph, an old roommate, across a chest-deep stretch to a small island in the river's center where we found the cabin we'd been hired to clean by its reclusive heiress owner.

Finally, there is the Hudson River. I spent many years living along the Hudson, first in Tarrytown, New York where I went to college and then on the Upper West Side of Manhattan where I occupied a series of illegal sublets during my early years as a journalist. I have also plied the Hudson on Circle Line cruises around Manhattan. But mostly the Hudson made my list because of one unforgettable image from September 1985. With Hurricane Gloria bearing down on the east coast, I walked from my apartment at the corner of Broadway and West 77th Street down to Riverside Park. I wanted to watch the river take the storm. I made my way through the park to the wrought iron railing, which runs along the promenade above the Hudson. A parade of tankers had taken refuge in the river. The giant cargo ships were lined up in the channel, single file, bow-to-stern, bow-to-stern, upstream and down for as far as I could see.

The Passaic River isn't on my list. But I began thinking about it several years ago after it tried to commandeer an essay I was writing about the American Southwest. Once the Passaic crashed my essay it was extremely reluctant to leave and I found myself struggling to whittle its presence down to a few sentences. I was supposed to be capturing the harsh beauty of parched desert plains and instead I was going on about floating trash and rusting bulkheads. I felt ambushed, like the river had been waiting all this time for a chance to hijack my attention. It worked.

When I finished the essay, I began researching the Passaic. I learned that the river, like the state it flows through, has a serious image problem. The Passaic is as historic as New York's storied Hudson and in some places— the 77-foot-high cascade in Paterson, for one— just as majestic.

But most people, including some New Jersey natives, have never heard of it. Those who have know it only as one of America's most polluted waterways. It's hard to bond with a river like that.

The Passaic is a poster child for rivers— for nature— everywhere. It was once the lifeblood of its region, source of food and power, playground of the rich, avenue of transportation, communication and commerce. The first white settlers sailed up the Passaic in 1662 and founded Newark, the nation's third oldest city, on its banks. The river's abundant charms fueled an explosion of growth and industry that helped transform the fledgling United States into a global manufacturing powerhouse. But in time the industrial revolution it helped to spawn would poison and betray the Passaic. By 1952, the year I was born, the river's beauty and majesty were dim and distant memories. Its lower stretch was a toxic canal. The Passaic wasn't a source of wonder and delight, or even interest anymore. For a generation, *my* generation, it inspired fear, revulsion and denial instead.

How did this happen? How could a river that was once celebrated for its beauty become a superfund site? How could my hometown river repel someone like me, a lifelong lover of rivers? And how did I wind up a river lover anyway after growing up along the sorry ass Passaic?

Could the river be rescued? Was anybody trying? Did anybody care, or was the Passaic just one more piece of leftover nature to be tossed out with the trash? What did nature mean to us anyway? And who would we be without it? Would the disappearance of the natural world, of its wild places and creatures, bruise our souls the way the loss of a loved one breaks our hearts?

The questions piled up. The answers, not so much. It was the summer of 2005. I started making calls.

The first person I reached was Andy Willner, who at the time was the Executive Director of the NY/NJ Baykeeper Association. Baykeeper is a nonprofit dedicated to protecting local water bodies. Andy was passionate, generous, cocky, fearless and a bit bombastic on the phone. I loved him. He called the NY/NJ Metropolitan Area a "big region" with "low environmental self-esteem." He said his mission was to awaken citizens to regional treasures like the Passaic. He said that people didn't know the Passaic anymore, that the river had become a stranger to them, and that you couldn't care about something you didn't know. He invited me to join him on a Passaic River boat ride.

I decided there on the phone that I might not be able to make people care about the river, but I could certainly reintroduce them, and myself to the Passaic. Surely there was more to my hometown river than the oily, garbage-strewn slough I remembered.

I've lived on the west coast for almost 30 years now, far away from my New Jersey roots. I strayed, in part, to escape environmental blights like the Passaic and the sadness they inspire. The Garden State can be hard on nature lovers. For all my time as a Jersey expat, though, I'm still a fiercely proud homey at heart. I've spent years defending my state from people who deride it as a giant toxic waste dump that every local calls "Joisey." It didn't seem like much of a leap to start defending my river too. The Passaic needed a champion. It had a story to tell, the story of an urban river at the start of the 21st century. I wanted to be the one to tell it.

As I began dredging up the river's past, the Passaic surprised me again. The river's story echoed my own: idyllic beginning, sad turns, lingering scars, uncertain future. The story got personal.

I'd been thinking about the project as my mea culpa to the river and to New Jersey, a way of saying I'm sorry for not sticking around to try to make things better. But maybe it was more than that. Maybe going back with the Passaic was a way to move forward with my own life. Maybe the river could help me make peace with the ambient sadness that seems, more and more, to haunt my present.

And so the book became a duet. Part natural history, part personal history, part adventure, part meditation on the wonder of rivers, the enduring ties of family, and the power of water and loss.

The Passaic River was calling me home.

1 | THE RIVER

MY MOTHER TOLD US NOT TO PLAY BY THE RIVER, and mostly we listened. But there were times when the riverside beckoned. Like those autumn afternoons when my older brother and I played football on the lawn behind the Homelite plant with our neighborhood pals.

The Homelite lawn was a natural football field, nearly twice as long as it was wide and almost perfectly flat. The grass was a thick, green cushion that was always neatly trimmed, not surprising really since lawnmowers were one of Homelite's premier product lines. There was no fence to bar our entry. No one ever chased us off. In fact, the Homelite workers in their dark blue coveralls would gather in the asphalt parking lot next to the field to smoke a Chesterfield or two and watch the game.

The parking lot formed the field's eastern boundary. On the other side of the grass and all along the western sideline was the Passaic River, a slick, dark menacing presence slinking its way down to Newark Bay. Across the river, the opposite bank had been replaced by a corrugated steel bulkhead. The dark armored wall, some 30 feet high, was punctured in places by huge round storm drains. The giant metal mouths, visible at low tide, drooled a frothy runoff into the river.

The bulkhead supported a mile-long, arrow-straight stretch of Route 21. Known locally as McCarter Highway, this particular section of road was a made-to-order speedway. Teenage hoods from Newark would gather there to drag race their GTOs. On summer nights, the whine and squeal of engines and brakes would drift across the river and seep into our dreams.

The Passaic is a tidal river by the time it reaches the Homelite plant. When the tide was in, the water would creep right up to the lip of the field. We never ran sweeps to the river side, always observing a five-foot buffer zone between our play and the water's edge.

There was another group of kids hanging around the Homelite field one afternoon. They were older kids, strangers to me. They were horsing around at the other end of the field. Kibitzing, my father would have called it. The Homelite workers eyed them suspiciously before ducking into their cars to head home.

When the parking lot had emptied and the sun had nearly set, another boy appeared. He was a husky boy, strong, with straight brown hair and roses in his cheeks. He walked across the grass on the diagonal, a path that took him between our game and the older boys' horseplay without disturbing either.

He was on his way home from football practice. He was wearing his uniform–Irish green jersey and white pants–and carrying his helmet by the face mask. The chinstrap made a ticking sound as it slapped against the helmet's plastic side.

I didn't hear what the older boys said to the football player. He paused, turned in their direction, then proceeded on his way, a little faster than before. They caught up with him at the far corner of the field, the river side. We had stopped our play by then, and stood nearby, in a huddle, watching.

The biggest boy shoved the football player in the chest. The football player shoved back. There was a scuffle and in the midst of it one of the other boys grabbed the helmet.

It was a beautiful helmet, snow white against the gathering dusk, no scuffs or grass stains, its top as smooth and round and shiny as a cue ball. The boys taunted the football player with their prize, yapping and hooting and prancing around him like jackals at a kill.

After a minute or two of this, the biggest boy, the one who started everything, stepped into the circle and claimed the helmet for himself. He raised it over his head and held it there. He locked eyes with the football player. Then, with a windmill motion of his arm, he hurled the white helmet into the Passaic.

It splashed in upside down, not 30 feet from where we all stood, gaping, near the water's edge. The helmet righted itself somehow and bobbed briefly, like a buoy, or a skull.

Frantic, the football player waded in after it. He skidded and slipped on the oil-slicked rocks. He sank to his ankles in the river's black sediment. Knee deep in the Passaic, its water staining his white pants a fecal brown, he tried to rescue the helmet with a tree branch. He managed to snag the facemask, but the river wouldn't let go.

Any one of us should have been able to dive right in and save that helmet. The water couldn't have been more than four feet deep where it went down. But we didn't. And the football player didn't either. No one dove into the Passaic River.

The Passaic wasn't fearsome in any traditional sense. It didn't rage or thunder. It didn't loll along and then suddenly turn into a boil or hurl itself over a cliff — not this far downstream anyway. It wasn't icy cold or booby trapped with eddies. It wasn't even that wide; a dog-paddler could make it all the way across. But the river scared us just the same.

We were afraid of its impenetrable darkness. Afraid of its industrial smell. We were afraid of the things that lived beneath its surface and the things that had died there. We were afraid of spotting a hand or a head bobbing in the rafts of garbage that floated by. We were afraid of submerged intake valves that sucked water into the factories along the banks. We were afraid of the river's filth. It wasn't the kind of filth that came from playing football with your friends. It was grownup filth. The kind that scared the blue out of water and coated the riverbank with oily black goo. It was the kind of filth you could taste; the kind that could make you sick, maybe even kill you. We were afraid of getting splashed with river water or of touching river rocks. We were afraid of falling in or—God forbid—going under. We were afraid of the river's anger at being so befouled, and afraid, most of all, of the revenge we felt certain the river would exact. The Passaic claimed the white helmet. It could claim us too.

It was 1960 in North Arlington, New Jersey, my hometown, a small borough on the Passaic's eastern shore just five miles upriver from Newark. This industrial lower stretch was our Passaic. My mother told us not to play by the river, but she didn't have to.

Many years later, in September 2005, I took my first cruise on the Passaic River with the NY/NJ Baykeeper Association, the New Jersey-based nonprofit that has taken up the river's cause. Our boat was a 16-foot Aqua Patio. It looked like a floating hot tub, all white with a high freeboard and banquette seating, ideal for the civilian river trips that the Baykeeper regularly runs up the Passaic. The two-hour tour took us about three miles upriver, from the mouth in south Newark to the New Jersey Performing Arts Center at the north end of downtown. It was the first time I had ever actually been out on the Passaic.

I took a seat in the bow with a pair of environmental engineers from Pennsylvania and three attorneys from the Rutgers Environmental Law Center. Janice and Martin, a retired couple from New York, were squeezed into the stern alongside two researchers from the New York Academy of Sciences, who were studying the ecology of New York Harbor.

Skipper Bill Sheehan, the Hackensack Riverkeeper, had the helm amidships. He was sturdy and gruff with a shark tooth necklace and a bushy red moustache. He leaned against the gunwale, just in front of Janice, one hand on the wheel. He had the look of a cop, or a bartender, or the ship's captain that he was—the look of someone who is comfortable being in charge.

Our host, Andy Willner, was a sunnier presence. It was the first time I had met Andy in person. He had a full gray beard and a thick shag of salt and pepper hair. A 35mm camera swung from his neck. He used his free hand—the one that wasn't gesticulating—to brace the camera against his middle-aged paunch. Andy had made this trip upriver on many, many occasions, but he snapped pictures with the eagerness of a first-timer, pointing out his favorite bridge and marveling aloud at the play of sunlight on the glass facades of the new office towers along the shore. Wonder lived next to outrage in his heart.

We set out from the Passaic Valley Sewerage Commissioners' (PVSC) massive sewage treatment plant on the shores of Newark Bay. Neat and surprisingly odorless, this 172-acre complex of circular tanks, pipes, pumps and stacks processes waste for 1.3 million residents in New Jersey's Passaic, Bergen, Essex and Hudson counties.

Once we cleared the dock, Andy unfurled a nautical chart and located our position in the labyrinth of bays, tidal inlets, islands and marsh. Raritan Bay was below us, linked to Newark Bay by the Arthur Kill, a tidal strait that separates New Jersey from Staten Island. Across Newark Bay to the east lay the so-called Meadowlands, the salt marsh that is home to the Hackensack River. Above us, and well within view, were the mouths of the Hackensack and the Passaic. The two rivers flow down from the north and squeeze the last bite of land between them into a chubby, muddy "V" called Kearny Point before they disappear into Newark Bay.

Andy straightened up, and with a sweep of his right arm, lassoed up the view. "All these bays were much larger," he said. "They were all extraordinary wetlands. The Passaic was one of the most bountiful rivers in the whole system, this estuarine stream with tributaries coming into it and a marsh system all around it."

I strained to picture the scene that Andy was describing. Like so much wild habitat in New Jersey, the wetlands that surround Newark Bay have been manhandled over time. In most places their transformation is so complete that discerning the natural features of the landscape is an exercise in extreme imagination. The once sinuous outline of Newark Bay, scalloped by coves and inlets and the mouths of its tidal rivers and creeks, is now ruler straight thanks to a century-long parade of large-scale public and private development projects. "You can see how geometric the shoreline is," said Andy, tapping the chart. "These are big fills."

The transformation of the Newark Meadows began in 1914 when the city of Newark, hungry for real estate, began reclaiming the marshland along the western shore of Newark Bay. Port Newark came first. The city dredged a mile-long shipping channel in the bay. They mixed the dredgings with garbage and ash and heaped the malodorous blend on top of the salt marsh until the landfill was firm enough to support the docks and warehouses that followed. By 1974, the Newark Meadows had completely disappeared, buried beneath the Port Newark/Elizabeth Marine Terminal, the Newark Liberty International Airport and the New Jersey Turnpike. Similar landfill operations soon claimed much of the eastern shore of Newark Bay too. Stands of white fuel storage tanks now occupy acres of former salt marsh in Bayonne. Welcome to the Garden State.

This massive industrial footprint is the first impression that most visitors to the state will have, certainly the millions who arrive and depart by way of Newark airport. And it's a lasting impression. The industrialization of the Newark Bay marshland has done more to stereotype New Jersey than all the jokes about big hair and the Mob. Newark Airport, Port Elizabeth, the New Jersey Turnpike and the Bayonne and Elizabeth fuel tanks are, alas, the icons of my home state.

My fellow Aqua Patio passengers seemed unfazed by the industrial sights and smells. Most were there on business. The environmental engineers were reconnoitering the Passaic for a client who had just bought riverfront property; the scientists were exploring the Passaic, Hackensack and Hudson River estuaries for a larger survey of New York Harbor; the lawyers were compiling an inventory of structures and businesses along the river. Janice and Martin were just looking for something interesting to do on a pleasant autumn afternoon. "Marty loves to be out on the water," Janice told me. The couple read about the Baykeeper tours in the newspaper, and drove out from their home in Manhattan.

They couldn't have picked a better day. The sky was a cloudless blue, the temperature a delightful 75 degrees. It was the kind of Indian summer evening that can make even the Passaic River look good. And it did look good. The water was actually blue. Its surface, miraculously free of debris, rippled and sparkled with every breeze. The sun was slipping lower in the sky. Three fingers from the horizon. Now two. The light was sharp and golden. We were sailing through honey.

We passed abandoned factories and rotting docks on the Newark side of the river, and a junkyard with towers of pancaked sedans, and acres of red and blue shipping containers stacked seven high. Backlit and spectral, each eyesore possessed its own odd beauty. They recalled a vanished era, the mid-19th century, when Newark was a king of U.S. manufacturing and the banks of the Passaic teemed with commerce.

About three miles upriver, just north of the Benjamin Moore paint factory, we came to the Diamond Shamrock superfund site. The address, 80 Lister Avenue, is on the far eastern edge of Newark, in the city's historic Ironbound neighborhood. Bill Sheehan maneuvered the boat in closer to shore, and shifted the engine into neutral. Most of the passengers stood—to take pictures, pay respects. Diamond isn't the only contaminated site along the Passaic, but it is by far the most notorious. For Passaic River advocates, 80 Lister Avenue is a battle cry.

From 1951 to 1983, the Diamond Shamrock plant manufactured pesticides and weed killers and close to a million gallons of Agent Orange, the defoliant that U.S. military aircraft sprayed onto the jungles of South Viet Nam during the war. The process of making Agent Orange generated huge quantities of dioxin, a poisonous by-product that remains the most carcinogenic substance known to man. Diamond's dioxin poisoned its workers, its plant site, the surrounding neighborhood and the river too. We had been right to be afraid of the Passaic.

The remains of the Diamond Shamrock plant had been entombed within the gray concrete mound we were floating past. It was the highlight of the tour. Roughly eight feet high and about the size of a football field, the mound was secured behind a concrete bulkhead and a steel fence, sealed with multiple layers of clay, and capped with concrete and an impermeable geofabric membrane. Within this waterproof six-acre grave lie the remains of the deconstructed Diamond factory buildings and 932 shipping containers filled with 66,000 cubic yards of dioxin-contaminated dirt, dust and debris that environmental cleanup crews literally vacuumed from the streets, stores, schools, houses, playgrounds and empty lots near the Lister Avenue property.

A few thousand years from now, said Bill Sheehan, archeologists studying this site will conclude that the people of the late 20th Century "built monuments to their pollution the way the ancient Egyptians built monuments to their pharaohs." With that, he kicked the engine back in gear and we continued slowly upstream. The skyline of downtown Newark was just ahead. Late afternoon sunlight lasered off the smoked glass facade of the FBI's new riverside tower.

"How come there are no other boats on the river?" asked Janice. Her face was hidden beneath the peak of her white cotton cap, which was pulled low against the harsh sun. It was a good question, direct and obvious, and it cut to the heart of things. Even the poison mound, and the Mad Max landscape and the occasional doomsday commentary from Andy and Bill hadn't managed to spoil the simple joy of being out on the water.

My mother would have liked this boat ride. She always dreamed of living by the water. Whenever she would mention this, my father would tease her: "You do!" he'd say. "You live by the Passaic."

There was a time when people would have coveted our home above the river. The Passaic was valued once, even beloved. Civic leaders harnessed its power to fuel their industrial revolution. Artists immortalized its beauty in paintings and verse. The river's clear, navigable waters sustained the settlers, who farmed and fished its fertile basin, and built cities and towns, like mine, along its banks. But those days didn't last.

The Passaic's beauty had been ravaged and its bounty spent long before Janice posed her question. No one in my large extended family ever mentioned, or seemed to mourn the river's passing. The Passaic was something we crossed over or drove along, but it was never something we engaged. The river was like an elephant in the living room of my childhood. Its death was a ho-hum fact of life, like Friday night shore traffic on the Garden State Parkway or Hudson County politicians on the take. Some people must have fought for the river once. But the battle was long over. The river lost.

How come there were no other boats on the Passaic River on this perfect late September afternoon? I knew the answer to Janice's question.

There are hundreds of thousands of waterways in the continental United States, 3.5 million miles of endlessly moving liquid. How many of these are technically rivers is a rather tricky question. "River" is not a scientific term. Indeed, science is a little *laissez-faire* when it comes to classifying a waterway as, say, a river versus a stream. It's not surprising then that rivers vary greatly in size and habit. Some are quite small, like the D River in Oregon, which flows 120 feet through Lincoln City to the Pacific Ocean. Some rivers are massive, like the Missouri, which at 2,450 miles is America's longest. Some rivers are ephemeral, surging into being after a desert downpour only to vanish with the rain, leaving behind nothing but a lacework of empty washes that hold the promise and threat of rushing water until the next big downpour. A few rivers, like Florida's Kissimmee, form gigantic puddles that sheet in slow motion, the gentlest flood inching across a grassy sea some 40 miles wide. Taken together, America's rivers drain the countryside like a giant open vascular system that collects water from the interiors of the continent and transports it to the seas. Their precious cargo is pirated along the way for drinking, bathing, irrigating, recreating and for powering millions of homes and industries.

Like the Passaic, most rivers are the *raison d'etre* for the communities and industries that have sprouted along their banks. There are thousands of river towns in the United States—Minneapolis, St. Louis, New Orleans, Augusta, Savannah, Albuquerque, El Paso, Cincinnati, Wheeling, Great Falls, Bismarck, Kansas City, Sioux City, Omaha, Trenton, Toledo, Fort Wayne, Wilmington. Those are just some of the larger ones. The Passaic spawned Newark (1666) and Paterson, New Jersey (1791), as well as dozens of smaller communities, and like most urban rivers, the Passaic has paid dearly for its largesse.

In strictly physical terms, the Passaic is a modest river, just 86.05 miles long[1]. Nevertheless, it is New Jersey's longest river, edging out the Raritan by a scant five miles. The name Passaic means "peaceful valley" in the language of the Lenni Lenape, the Native American tribe that occupied northern New Jersey before white settlers arrived.

The river rises in Mendham, a historic township in north central Jersey. It heads almost due south at first, then veers sharply north, then northeast, then due east and then south again, making two final northward loops before emptying into Newark Bay. This erratic path traces a sloppy, upside down U that winds through, over, under, and around five New Jersey counties, some 40 of its cities and towns, three swamps, four dams, four meadows, four waterfalls, a pond, a lake, 53 bridges and seven highways, and past countless homes, parks, playing fields, parking lots, diners, junkyards, office buildings, shopping centers, gas stations, warehouses and factories. The drive from Mendham to Newark is about 30 miles. The Passaic takes the long way around.

Through its 86-mile course, the Passaic is many rivers: swift and clear in its upper stretch, sluggish and swampy in midsection, a thundering cascade at Great Falls, brackish below the Dundee Dam, and so industrial

---

1 U.S. Geological Survey, West Trenton, New Jersey, 2011.

in its final miles that New Jersey poet laureate William Carlos Williams declared it "the vilest swillhole in Christendom."[2]

The river can be divided into three long stretches. The Upper Passaic is a largely downhill romp through meadows and forest along the southeastern edge of the Great Swamp National Wildlife Refuge. The Central Basin is the long, flat, flood-prone middle reach that flows north for some 40 miles through an ancient lakebed. The Lower Valley, where I grew up, is a 35-mile-long corridor that sweeps down from the cliffs of Paterson to the sea level marshes of Newark.

This convoluted journey from pristine headwaters to superfund site mirrors the triumphant and tragic relationship between nature and industry in America. The wildness and beauty that awed the first settlers some 400 years ago eventually powered the mills, farms and factories that produced clothes, food, steel and electricity, a robust international trade and a large and solid middle class. But along the way, the mighty frontier that helped forge American enterprise and character fell victim to an industrial fervor that seemed, at every turn, to sacrifice natural resources for financial gain. The power and much of the breathtaking natural beauty of our national rivers survives today only in isolated patches, and then just barely. "Our tools are better than we are," wrote the late naturalist Aldo Leopold in his 1949 environmental classic *A Sand County Almanac.* "They suffice to crack the atom, to command the tides. But they do not suffice for the oldest task in human history: to live on a piece of land without spoiling it." As my great grandmother, Emily Sullivan, liked to say: "Don't shit in the nest." The Passaic River is an object lesson in what can happen when we ignore that simple, salty advice.

---

2 From *In the American Grain* by William Carlos Williams, 1956.

The Passaic changes character in the Lower Valley portion of its run. Above its mouth in Newark Bay, the Dundee Dam crosses the river. The Passaic is freshwater above the dam. Below, the river becomes a swirl of freshwater and seawater whose salinity varies with conditions of weather, river flow and ocean tide. Water levels in the river fluctuate about five feet with each daily tide. During extreme high tides, the Passaic can rise as much as eleven feet. When conditions are right—a high tide during the dry summer season, for instance—the tongue of saltwater from Newark Bay can lick the Dundee Dam, a full 17 miles upstream.

The tide was going out on the day the football player lost his helmet. I remember the river receding as we played on the Homelite field that afternoon. I imagine the white helmet drifting down through the lighter layer of fresh water, then farther still through the heavier salt layer, until it reached the river's mucky bottom. It may be there still, buried like dioxin beneath two generations of mud and silt. Or maybe the tide took the helmet with it, out to sea.

The Aqua Patio passengers were all quieter on the return trip, even Bill and Andy. I wondered what everyone would take away from this experience. Andy used the Passaic River cruises to shake people up, open their eyes, confront them with the tragedy and the possibility of the Passaic. Later that summer, he would take the mayors of Newark and Harrison out on the river. Baykeeper hosts cruises for local business leaders, for the press and for the general public too.

"Our job is to make advocates of people," said Andy. He was giving me a lift back to my car, steering his Subaru Outback slowly along the paved streets that wind through the Passaic Valley Sewerage Commissioners' plant from the riverside dock to the parking lot at the main entrance. "Remember *Moby Dick*?" he asked, out of the blue. "The first chapter is all about Manhattan. When industry and pollution kind of took the water away from people, the people responded appropriately: they turned their backs on the waterway and took on other interests. Same thing with the Passaic. When the Passaic became foul, when it was no longer a place to picnic and boat and swim, it became less known to everyone except the people who worked on it. And those people used it as a highway and a toilet, and when it started to smell bad and people started to hear warnings about it, the Passaic became an unknown place."

I left Andy standing in the parking lot, deep in conversation with the two environmental engineers from the cruise. I tooted the horn and waved to them as I passed through the gates and out of the plant. Then I fished my rumpled directions from the glove compartment and followed them backwards to my brother Paul's house in Cranford, where I always crash when I'm visiting New Jersey. We call it the Cranford Hilton.

My maiden voyage on the Passaic River had the desired effect. Andy would have been pleased. I didn't get over my fear of the river. But after the boat ride that fear mingled with curiosity and a kind of compassion. The river had touched me. Look at me, it seemed to say. Listen.

The lower Passaic has been unknown and unloved for more than a century now. That's a long, sad estrangement from the two generations of New Jerseyites who live on and near its shores. But unknown is not unknowable. Unloved is not unlovable. Driving back to Cranford that evening, I decided to kayak down the Passaic. Paddling its length seemed like a good way to get reacquainted. It was time I got to know my hometown river.

2 | NEWARK

THE STORY OF THE PASSAIC is inseparable from the story of New Jersey's oldest city, and so I'll begin the account of my Passaic River voyage at the end— of the river, that is— in Newark.

We paddle in with the dusk, Carl, Cathy and I, around a bend in the river and past a decommissioned bascule bridge frozen in the open position with a welcome message spray painted on its underside: Newark Sucks.

The river is narrow on the approach, almost a culvert, corseted by bulkheads on both banks and domed by closely packed, rusting iron bridges. Newark's first bridge across the Passaic was a wooden contraption that spanned the river at Bridge Street. Its completion in 1795 sped the transport of people and goods between Newark and New York City. Today, seven bridges carry countless cars and trucks and buses and trains and lots and lots of people in and out of Newark.

The modest skyscrapers of downtown rise up from the western shore. On the opposite bank, in Harrison, light industrial buildings sit in a field of tawny grass. The grass evokes the salt marsh that once enveloped the lower Passaic. But the grass won't survive much longer. The Harrison side of the river is destined for major development. Red Bull Park, a soccer stadium and the first phase of the Harrison Metro Centre project, will replace the open spaces and the lovely old Charles F. Guyon Pipe Fittings and Valves factory. The massive, barn-like structure, which loomed so gracefully over its neighbors, is being dismantled board by board. On the evening we paddled past, the factory had already been flayed. Its intricate wood framing was exposed but intact, a graceful skeleton against the lavender sky.

The water before us is dead calm, an impenetrable obsidian mirror. It is Sunday evening. October. In a few minutes streetlights will begin to flicker on, but for now the city is hushed and shrouded in twilight. A lone pedestrian, a teenage boy, makes his way across the first of two decrepit iron bridges ahead. His face is hidden deep within the folds of a dark blue hoody. I wave, but he doesn't respond.

I slow down as I approach the rusting NJ Transit Bridge, and start to back paddle when I notice a train pulling out of Penn Station. I don't want to be under the bridge when the train crosses it. I can just see the headline: CRUMBLING BRIDGE CRUSHES KAYAKER. I lay my paddle across the cockpit and cover my ears against the squeal and rattle of the train wheels on the steel tracks above.

Ten feet below me, buried in the dark pudding of silt and muck at the bottom of the river, is one of the world's largest deposits of dioxin. Andy Willner once told me that the sediments in this stretch of river are so richly contaminated that "if dioxin had a use you could mine it." There are other poisons down there too: polychlorinated biphenyls (PCBs) and polycyclic aromatic hydrocarbons (PAHs) and dichlor-diphenyl-trichlorethylene (DDT) and the whole cast of heavy metals—copper, cadmium, chromium, lead, zinc, nickel, mercury. All the nasty by-products of Newark's electroplating, hat-, paint-, varnish-, leather-, fertilizer- and pesticide-producing past are there below me, bound up in the Passaic

River mud. Dioxin is the most deadly of them all, the most carcinogenic substance known to man. I draw my paddle up out of the water and slide the small rubber drip guards as close to the paddle blades as they'll go. I don't want any of this water dribbling down the paddle shaft and dripping on to me.

The highest sediment concentrations of dioxin are just downstream at the site of the old Diamond Shamrock chemical plant. But the Passaic is a tidal river here. Opposing forces of ocean tide and river current have been volleying toxic particles up and down this 17-mile stretch of Passaic, from the Dundee Dam to Newark Bay and back, day after day, decade upon decade for the last 50 years. From Newark Bay, the Passaic toxins have found their way into Arthur Kill and Kill Van Kull, the tidal straits between New Jersey and Staten Island. They have drifted as far as New York Harbor.

Every Passaic River flood—and there have been many—sends river water and its sediment load surging horizontally out across the river's floodplain and into the marinas and industrial parks and playgrounds and ball fields and towns where boats and trucks and people carry the toxins farther still. No one really knows how far the toxins have traveled or where they have eventually wound up. Traces of dioxin and other poisons may still linger on bulkheads and hulls and pilings and tree trunks and any other surface within the river's long watery reach.

In January 2012, four months after Hurricane Irene and Tropical Storm Lee battered the east coast, federal environmental investigators found high levels of dioxin in water samples taken from the Passaic near Riverside County Park in Lyndhurst, New Jersey. Lyndhurst is a few miles upstream from the Diamond Superfund site. The discovery is a wake up call, the first documented evidence of recent dioxin migration. "The perception has long been that the pollution is down in Newark, that we don't need to worry about it," NY/NJ Baykeeper's Executive Director Debbie Mans told the *Bergen Record*, "when in fact it's in your backyard."

The irony of dioxin is that no one set out to make it. Though it was (briefly) auditioned as a flame retardant for wood, dioxin has no industrial purpose and no inherent value, economic or otherwise. The dioxin in the Passaic River was an accident—at first. A useless by-product created during the process of making the infamous herbicide Agent Orange. Call it sloppy chemistry. Then workers at a German chemical factory began getting sick, and German chemists began refining their manufacturing process. The modifications they made reduced dioxin levels to near zero. Some chemical companies adopted the newer, safer protocols. For those that didn't, the continued production of dioxin would become a crime.

Dioxin is a celebrity toxin. Love Canal, New York. Times Beach, Missouri. Seveso, Italy. South Viet Nam. Newark, New Jersey. Dioxin was the star in each of these notorious environmental disasters. Dioxin made headlines again in 2004 when it was used to poison Ukraine president Viktor Yushchenko. It turned Yushchenko's once handsome face into a lumpy, grayish, mottled mess.

There are dozens of scientific papers and reports and books and web sites devoted to dioxin. We know exactly what dioxin looks like and where it comes from. We know how it behaves in the environment and what it can do inside the human body. We know a lot about dioxin, and everything we know about it is bad.

The U.S. government regulates industrial emissions now. Federal restrictions have significantly reduced the levels of dioxin that are released into the environment. But we still generate dioxins every day with our coal fired power plants and waste incinerators and pulp mills and metal smelters and diesel truck exhausts and landfill fires and plain old backyard trash burning. Once this poison gets into our air and water and fields and plants and animals and us, we have no idea how to get rid of it.

Dioxin is a nickname for polychlorinated dibenzo para dioxin. It belongs to a large extended family of compounds, more than 400 at last count. Family members look alike. Each dioxin molecule has two, six-sided rings of carbon atoms. The rings are linked like train cars by one or two oxygen atoms, and dotted with chlorines. It's the chlorine atoms—their number and location—that make dioxins so dangerous. According to the U.S. Environmental Protection Agency, there is no "safe" level of exposure. But some dioxins are more toxic than others. The most toxic by far is TCDD, 2,3,7,8-tetrachlorodibenzo-para-dioxin. TCDD is the yardstick against which the toxicity of all other dioxins is measured. TCDD is the dioxin buried in the sediments of the Passaic.

"It's an amazingly potent carcinogen," says Robert Chant, Associate Professor of Physical Oceanography at New Jersey's Rutgers University. Chant has been modeling the movement of dioxin-contaminated sediments in the Passaic. He says there are only a few places on earth with higher concentrations of dioxin. One is the jungles of South Viet Nam where, from 1962 until 1971, America's Operation Ranch Hand sprayed 19 million gallons of Agent Orange. A hand-painted sign hung above the door in the staging room at the Saigon airport where U.S. Air Force pilots prepared to fly their Agent Orange sorties. It said: "Only you can prevent forests."

When Chant wants to impress his undergraduate students with dioxin's singular toxicity he gives them his former-Secretary-of-State-Colin-Powell-testifying-before-the-United-Nations routine. "I hold up a vile with four grams of sediment, falsely claiming that it's from the Passaic," says Chant, "and explain that the vial contains enough dioxin to contaminate a lifetime's supply of drinking water."

Four grams is about three-quarters of a teaspoon.

We haul our kayaks out of the Passaic at Riverbank Park in the Ironbound section of Newark. The river's edge is low and wide here, and flat—a perfect take-out spot. The Ironbound is a four-square-mile neighborhood in Newark's East Ward. It's a tight, working class community where factories front the river and row houses with aluminum siding squeeze into the narrow streets behind. The Ironbound lies due east of downtown Newark along a smooth northward loop in the Passaic River. The loop reminded locals of a long slender neck, hence the neighborhood's original nickname, "Down Neck," which is what old-timers still call the area. The fictional Tony Soprano grew up Down Neck.

There are two theories about the genesis of the neighborhood's more recent "Ironbound" handle. It could be a nod to the many forges and foundries that operated in this part of the city during the last half of the 19th Century. Or it may reflect the fact that the neighborhood is surrounded by railroad tracks. With the steel ties of NJ Transit to its north, and Conrail to its west and south, the Ironbound is literally iron bound. One hundred of New Jersey's most toxic sites lie within this steely embrace. The Ironbound zip code, 07015, is the most contaminated in the state.

Riverbank Park is the only place in the Ironbound where you can actually get to the Passaic. We slalom through some rotted old pilings whose ragged tops poke out of the water, and crunch ashore on a rocky muddy beach. I have a photo of Carl, my river guide, amidst the pilings. He is still sitting in his beached blue boat, intently focused on taking a picture of the now-twinkling Newark skyline. His navy blue watch cap is pulled down low against the chill. The reflective wristbands on his red slicker catch the camera's flash, giving Carl two bracelets of glowing white light. Wrist halos. Behind him, a few hundred yards or so downstream, acres of neatly stacked shipping containers form the horizon, a broad red and green stripe between the deep blue of the river and the fading blue of the sky. We are about three miles from the river's mouth.

New Jersey's first white settlers landed just south of here in May 1666. They were an advance guard of Puritans from Connecticut's New Haven Colony who sailed down from New England hoping to find a place where they could insulate themselves from society's corrupting influences. Historians aren't certain about the exact date of their arrival. No letters or diaries of the trip survived. But consensus has formed around May 18th. In *Newark*, his 1966 history of the city, New Jersey historian John T. Cunningham imagines the final leg of the journey as their small fleet sailed across the great expanse of Newark Bay and tacked north into the wide blue mouth of the Passaic. The voyagers would have been completely surrounded by salt marsh, a rippling carpet of tall, spring-green grass that fanned out in every direction. Its edges were trimmed "blue with iris" and in the forested hills beyond, "white dogwood brightened the woodland."

There were 30 large families in that first party, including Samuel Swain, his 17-year-old daughter Elizabeth, and her fiancée Josiah Ward. Elisabeth and Josiah were the first ashore. A 40-foot mural commemorating this moment hangs in the Essex County Courthouse. "Two young lovers, then," writes Cunningham, "were the first settlers to put foot on soil that would give rise to a great city."

More settlers would follow. Sixty four families in all. Catlings and Camfields, Wheelers and Tichenors, Burwells, Bruens, Brownes, Kitchells, Cranes, Freemans, Lyons, Peckes and Penningtons, Piersons and Baldwins; the surveyor Edward Ball; and the brothers Albers and Hugh Roberts, who were accomplished leather tanners. They came from Branford and Milford and New Haven and Guilford, nearly 400 of them, all eager to build a new home.

Their captain was Robert Treat. He was 41 years old, a respected leader of the New Haven Colony, and the father of eight whose great grandson Thomas Treat Paine would sign the Declaration of Independence. Newark's homage to Treat, its founder, is a bronze plaque on the building that now stands at the site of his old Market Street home, and an eponymous 15-story hotel on Park Place. My parents held their wedding reception at the Robert Treat Hotel on June 10, 1950—284 years after Treat led his band of seekers up the Passaic.

It took some negotiating with chiefs of the local Hackensack and Lenni Lenape tribes, but Treat managed to procure most of what is now Essex County. He paid the natives with rifles, pistols, gunpowder, lead, swords, knives, beer, blankets, britches, hoes, axes, kettles and "three trooper Coates." Everyone went home happy.

The territory that Treat bartered for extended west from the Passaic to the foot of the first Watchung Mountain Range. In time, it would be partitioned into the towns of Belleville, Nutley, Bloomfield, Montclair, Glen Ridge, the Oranges, Irvington, sections of Maplewood and Short Hills and, of course, the city of Newark. But in the beginning there was just the land and its bounty. The eastern section was a sprawling salt marsh riddled by streams that flowed down from the Watchungs and into the Passaic. Fresh water was plentiful. Fish and game were abundant. There were stands of cedar in the uplands. Farther west, the Watchungs were thick with native hardwoods that the new settlers would harvest to build their city on a plateau overlooking the river. The Connecticut Puritans had found their promised land, and they set about remaking it.

Town fathers quickly laid out a grid for their city and held a lottery to assign properties to the founding families. In recognition of his fine leadership, the grateful citizens of Newark gave Robert Treat extra acreage and first pick. Treat took two parcels, eight acres in all along Market Street, the new city's main boulevard. But Treat did not stay in Newark. He went back to Connecticut in 1672, eventually serving as that state's governor. By the time he departed, homes had already begun filling Newark's new streets, and the new community had raised its own Meeting House. "Once housed, clothed and lifted in soul and spirit by the church," writes Cunningham, "the Puritans turned to other fundamental concerns."

One of those concerns was "the meadows," the name given to the surrounding marshlands. From the beginning, the marsh was seen as an obstacle. According to Cunningham, "[C]onstant effort was expended to make the great morass accessible and useful." The soggy prairies—half mud, half water—were too wet to build houses on or easily lay roads across. Each spring they gave rise to dark swarms of mosquitoes. The marsh needed fixing. Colonists sent about draining some sections and burning others. They built dikes. They added fill. They did what was necessary to make the marsh more suitable for building and farming and grazing—more suitable for human habitation and enterprise.

The marsh did have one benefit. It was a good source of salt hay, a species of cord grass (*Spartina patens*) whose delicate, hollow stems sway with the wind and waves, and give the marsh its undulating surface. The settlers harvested salt hay to use as food and bedding for their farm animals. Beyond that, the newcomers didn't seem to have much use for the marsh. Nor did they seem to recognize the pivotal role it played in the ecology of their new home. How the marsh recycled nutrients and purified the water. How it provided shelter and a spawning site for local fish and shellfish, and a hunting ground for the mammals and birds that were their natural predators. How the marsh absorbed the storm tides and spring floodwaters that would rise up to threaten everything they held dear.

If the Puritan settlers did grasp the workings of the marsh, if they did appreciate its value and honor its place in God's creation, perhaps they rationalized its alteration. There was so much of it, after all, acres and acres in every direction. The Newark "meadows" was just one small sliver in a vast sea of salt grass that, at the time, fringed the entire east coast. The settlers had sailed past a good stretch of it on their voyage from Connecticut. The marsh landscape was familiar to them, even comforting. Samuel Swain exploited this fact. When he returned to Connecticut to lure more families to the new "earthly Eden" to the south, he assured the hesitant that the green wetlands of Newark would remind them of their Connecticut home. The original settlers and all who followed would forever underestimate or ignore the effect of their collective industry on the delicate balance of the marsh.

Within 100 years of their arrival, they would erect their city and lay a crude east-west roadway across the marsh. The road was the first direct and mostly overland link between Newark and Paulus Hook, the site, in present-day Jersey City, where ferries shuttled people and goods across the Hudson to New York. The overland portion of the trip to Paulus Hook was interrupted twice when travelers and cargo had to be ferried across the Passaic and Hackensack rivers. By 1795, drawbridges across both rivers and a wider roadway greatly shortened the travel times, which helped make Newark a vital hub for travelers and trade.

New Jersey's Meadowlands are what's left of the vast cloak of wetlands that once wrapped the lower reaches of the Passaic and Hackensack Rivers and Newark Bay.[1] They are part of the great estuarine ecosystem that also includes the Hudson River. Some 8,400 acres of salt marsh remain,

---

1 The 32-square-mile area, which includes portions of 14 cities and towns in New Jersey's Hudson and Bergen counties, is now regulated by the Hackensack Meadowlands Development Commission. Established in 1969, the Meadowlands Commission attempts to strike a balance between "preservation of the delicate balance of nature" and "orderly, comprehensive development."

bordered by Jersey City, Secaucus and North Bergen on the east; by Kearny, North Arlington, Lyndhurst, Rutherford, East Rutherford and Carlstadt on the west; and to the north, by Little Ferry, Moonachie, Ridgefield, North Bergen and Teterboro. The Meadowlands have always been a dumping ground—for sewage, garbage, and all manner of hazardous and haphazard waste. The marble remains of New York's old Pennsylvania Station are thought to be scattered out there amongst the reeds, along with rubble from the World War II bombing of London—it provided ballast for returning warships—and more than a few victims of foul play. A 1969 survey by the state determined that 118 New Jersey communities were piling more than 1.5 million tons of trash onto the Meadowlands every year.[2]

New Jersey State Route 7 is one of two public roads that cross the Meadowlands, west to east. (Route 3 is the other.) Route 7 is about seven miles north and roughly parallel to that first crude road constructed between Newark and Paulus Hook. Known locally as the Belleville Turnpike, Route 7 is a straight two-lane roadway that begins on the eastern bank of the Passaic, at the Rutgers Street Bridge in North Arlington, and ends eight miles later at the mouth of the Holland Tunnel.

This was the route we took to Jersey City, hometown of my father's large Italian family. Every Sunday we'd make the drive into Jersey City to visit my father's maternal grandmother, the family matriarch. We called her Grandma Larch because she lived on Larch Avenue, a treeless dead-end block of dollhouse homes tucked under a Route 7 overpass, just a block or two from the American Can Company. Getting there from North Arlington was a quick 10-minute car ride due east along Route 7.

The roadway's surface has been widened and repaved on the approach to Jersey City, but otherwise Route 7 is pretty much the same as I remember it from those Sunday drives: a bumpy, potholed, lonely strip that comes

2 Source: New Jersey Meadowlands Commission (www.njmeadowlands.gov/index.html)

down off the North Arlington ridge, across the flat swampy Meadowlands, over the Hackensack River and into Jersey City, to Grandma Larch's house.

Driving along Route 7 at night could be unnerving. Driving through in daylight was disgusting and strangely beautiful. Fetid water with an oily surface sheen filled the gullies alongside the road. Cracked, leaky pipes, old tires and assorted lumber and trash lay partly submerged in the shallow roadside pools. But now and then, we'd catch site of a long-legged wading bird, some heroic—or foolhardy—heron or egret. Proof of life.

There were other sights along the way to Grandma Larch's. At the foot of the ridge we'd pass the North Arlington landfill. Today, the mound is covered over and planted with grass. Back then it was a raw, exposed mountain of trash. Even on Sundays, trucks would chug up its fragrant flanks with an entourage of greedy gulls swooping and bickering overhead.

A little farther along Route 7 sat the radio tower of WMCA, New York's most popular AM station. From its New York studio, deejays Dan Ingram, Harry Harrison and "Cousin Brucie" beamed the soundtrack of the '60s. The station's Meadowland outpost consisted of a tall radio tower with its blinking red light and the squat cinderblock building at its foot. The two structures were stranded side by side on a little island in an open water stretch of marsh. A narrow dike linked the island to Route 7. Returning home after dark I sometimes saw a light on in the cinderblock bunker. It made me wonder whether someone actually worked in there. If they did, I thought, it must be the loneliest job in the world.

Just before we reached Larch Avenue, we'd pass Arco Welding & Machine Works, two low red brick factory buildings surrounded by a hurricane fence and dwarfed by a nearby stand of round, white fuel storage tanks. My father always pointed out the buildings as we sped by. Arco was his father's business. A producer of bridge beams, highway joints and the storm drains for the George Washington Bridge, Arco boomed during

World War II, and nosedived shortly after. It was finally shuttered in 1959, five years after my grandfather died. Declaring eminent domain, the state bought the buildings and the land in 2000 as part of its plan to widen Route 7. My grandfather's business is now buried beneath the asphalt.

The road from Newark to Paulus Hook ushered in big changes for the city of Newark. Before it was built, Newark had been a remote outpost. Once the new road opened, the city became a pit stop along the well-traveled route between Philadelphia and New York. Newark's modest shoe trade with the South, which had started in 1790, began to flourish. Its success was opening newer, more far-flung markets for other Newark-made goods. By the close of the 18th Century, Newark's well-made shoes and well-traveled highway raised the city's profile and accelerated its transition from a small colony of religious refugees desperate for solitude into a bustling city eager to do business with the outside world. Nineteenth-century Newark would greedily embrace the society that its 17th-century founders had traveled so far to escape.

By 1860, Newark's population had topped 70,000— up from just 11,000 in 1830—and manufacturing was the engine that drove the city's growth. Newark possessed all the raw materials for manufacturing success. It enjoyed a ready supply of water to power its industry, a navigable river and new roadways to transport its goods to market, and perhaps best of all, a diligent and creative workforce. Many of the city's skilled tradesmen were one-time farmers, who had taken up leatherworking or some other craft as a way to make a living during the fallow winter months. They excelled at improvising solutions to even the most intractable industrial problems. The city's surfeit of "clever hands," writes Cunningham, attracted entrepreneurs and inventors, who tackled the challenges of the day with ingenuity and alacrity. "Newark's industrial power lay in its ability to meet the needs of an altered world," observes Cunningham. "Opportunity often lay simply in asking, 'what must people have?'"

As Cunningham tells us, when the city's furniture makers needed a sealant to protect their handmade wooden pieces, Newark's Samuel P. Smith gave them the first clear non-sticky varnish (in 1835). When Newark's farmers wanted to improve crop yields, Joseph and Alfred Lister, newly-arrived from England (in 1850), built a factory on the Passaic that turned livestock bones into fertilizer. When city builders cried out for a white house paint that could stand up to the elements, the New Jersey Zinc Company opened a plant in Newark and developed the first white zinc oxide for paint.

Newark was brimming with innovation.

Edward Balbach, a German immigrant, built a refinery on River Street in 1850 and using scraps of discarded gold and silver salvaged from the floors of Newark's many jewelry factories developed revolutionary techniques for separating precious metals such as gold and silver from their baser cousins such as nickel and lead.

Thomas Edison arrived in Newark from New York City in 1871 and spent the next five years recruiting workers with "skilled fingers" for his small Newark laboratory and for the larger "invention factory" he would start up in Menlo Park, New Jersey in 1876. Edison's Menlo Park lab produced the first light bulb and phonograph.

Edward Weston, native of England, electrical genius and Edison rival, came to Newark in 1870. By 1882, Weston was lighting the city's streets with his new electric arc lamps. He used the same newfangled lights to illuminate the Brooklyn Bridge— and bedazzle the crowds— at the bridge's 1883 opening. As founder of the Newark Electric Power & Light Company, Weston patented numerous electrical instruments and measurement devices. But his crowning achievement was the invention of the Normal Cell, the standard used to measure the volt.

Seth Boyden was the most versatile and selfless Newark inventor of them all. Convinced that the city's tanning industry could make good use of his latest invention, Boyden moved from Massachusetts to Newark in 1815. His machine for shaving leather into thin sheets became an instant hit. Boyden went on to invent patent leather and malleable iron. He developed a new device for making nails, and new methods for silver-plating belt buckles and reducing zinc ores. He built his own steam engine. He invented a machine that formed hat bodies. He developed a new strain of strawberries. About the only thing Boyden didn't do was patent his inventions, save for one request filed near the end of his life for the hat-making machine. Boyden believed in sharing his work. He died impoverished in 1870 at age 82.

Seth Boyden is commemorated by a bronze statue in Newark's Washington Park. He stands in front of an anvil, wearing his long leather apron. When the statue was unveiled in 1890 it was the first to honor an American workingman. The plaque on the pedestal lists Boyden's inventions in chronological order, and praises him as "a mechanical genius: a philosopher: a modest, helpful citizen of Newark."

The list of Newark inventors and entrepreneurs goes on. For me, a New Jersey native, who was born in Newark, at St. Michael's Hospital, and raised just north of the city, the most striking thing about these historic achievements was my utter ignorance of them. In all the history lessons I sat through as a student at Queen of Peace elementary and high schools in North Arlington, I never remember studying the story of my state.

I met Charles Cummings in May 2006 on my first research trip to the Newark Public Library. Cummings grew up in the south, but Newark was his adopted home. He was the city's official historian, overseeing the library's New Jersey Information Center, a trove of city history.

Tall and lean with thinning gray hair, Cummings wore a gray tweed suit with a white shirt and bow tie. He managed to look both natty and a bit rumpled, a fading southern gentleman. His gait was a little stiff that morning as he came out from behind the information desk to greet me. He hadn't been feeling well of late.

We sat at one of the long wooden library tables in the New Jersey Information Center's third floor space. Cummings leaned back in his chair and spun me a ten-minute tale of Newark's industrial years, adjusting the pace of his narrative to match the speed of my scribbling. He spoke of the early industries, the saddle and harness makers and the brownstone quarry that supplied Manhattan builders. He noted the importance of shoemaking and how by 1806 or so fully one-third of Newark's workers were employed in the growing shoe trade. He described the waves of immigrant labor: the Germans, Irish, Italians and Jews in the 19th Century; and in the 20th, the African-Americans from the South, American Africans, mostly from Nigeria, Hispanics from Cuba and Puerto Rico, and the most recent arrivals from South and Central America. The Civil War, he said, brought "great prosperity" to the city. Newark factories turned out many items for the Union army, including boots, caps and waterproof coats. He singled out the Scottish industrialist George A. Clark, who by 1870 employed hundreds of men and women in his sprawling riverside thread mill and "treated workers very well." He said that by 1900 pollution had finally put an end to the annual Passaic River rowing regatta.

Cummings lit up when he talked about Newark. The fact that I'd be able to find detailed floor plans of Newark's historic factories in the city's Fire Insurance Atlases genuinely thrilled him. The whole subject of the city put a twinkle in his eyes, as if he was actually seeing the events he was describing, watching the historic city in its heyday.

I have my own memories of Newark. Driving through Branch Brook Park each year at Easter time to see the cherry blossoms. Visiting my father at St. Michael's Hospital. Christmas shopping at Hahne & Company, Newark's oldest and most urbane department store.

In 1974, a year or so after my father died, my mother took a job at Hahne's, in the men's department. It was a big step for her. She hadn't worked outside the home for more than 20 years, not since 1950 when she got married and left nursing. But her job at Hahne's kind of broke the ice. A year or so later, she screwed up her courage, took the required three-month refresher course and went back to nursing. She got a job at Newark's Columbus Hospital in 1976.

Columbus is in a section of Newark called the North Ward, a densely packed residential neighborhood, north and west of downtown, that includes Branch Brook Park. The hospital is actually closer to Bloomfield than it is to downtown Newark, which gave us kids some measure of comfort because my mother worked the 3-11pm shift. She was the charge nurse on 2 North, a general medical/surgical floor that would be filled with AIDS patients by the time she retired in 1988. In 1985, my mother was Columbus Hospital's Nurse of the Year.

The Newark Library's Main Branch is at 5 Washington Street, well south and east of Branch Brook Park and just a few short blocks from where Hahne's used to be. (The company closed its Newark store in 1986.) The library is a dignified three-story structure, built of granite and marble in the Italian Renaissance style with large arched windows and a copper roof. It opened in 1903.

I first visited the Newark Library as a young teenager in 1965. I went there looking for a piece of sheet music. The Newark Library had an extensive sheet music collection, which was housed at the time on the very top floor. The day I spoke with Charles Cummings was the first time I'd been back to the Newark Library in all those many intervening years.

When he finished his cogent overview of Newark's industrial history, Cummings laid out the path I would follow in fleshing out the details: first, John T. Cunningham's history of the city, then Peter J. Leary's industrial history of Newark, then Stanley Winters' compendium of oral histories followed by the Newark *Industrial Directory, 1836-1962*, William H. Shaw's *History of Essex and Hudson Counties, New Jersey*, the early 19th Century Shoemaker map— named for its shoemaker cartouche— the 1868 Van Dyne atlas, and the directories of Benjamin Thompson Pierson, which were published annually from 1835 to 1862 and in which Pierson tracks Newark's transformation from sleepy farm town to industrial powerhouse with census-like precision.

Cummings instructed one of the younger librarians to assemble the research materials for me. Moments later, there was a stack of musty oversized books on the table in front of me. Charles Cummings wished me luck and disappeared into the large room behind the information counter. That was the last I saw of him. He died eight months later on December 22, from complications following heart surgery. He was 68. Since his death, the room where we met, and which houses the library's extensive New Jersey collection, has been renamed. It is now the Charles F. Cummings New Jersey Information Center. An oil portrait of Cummings, wearing his gray suit and bowtie, hangs high up on the wall across from the information desk above shelves of local newspapers, magazines and periodicals.

In any reading of Newark's industrial history one thing is clear: 19th century Newark was a city of problem solvers. Clever industriousness gave rise to many novel products and manufacturing processes, a good many patents, and a great deal of wealth. It also produced an awful lot of waste, which might not have been so troubling except for the toxic nature of Newark's industries. The city's factories turned out paints and varnish, hats, jewelry and leather goods. In time, Newark plants would also make herbicides, pesticides and petrochemicals. From its earliest days the city was home to some of the dirtiest industries on the planet. Most of Newark's factories sat on the banks of the Passaic. Much of the waste they generated was toxic, and most of it eventually wound up in the river.

Dumping industrial and human tailings into a nearby river or creek was a common method of waste disposal in 19th Century America. At the time, it made perfect sense. Current whisked the waste downstream and out of sight while resident aquatic flora and fauna got to work breaking it down and down and down again until all that remained of the garbage was an elemental broth of carbon, hydrogen, oxygen and nitrogen atoms ready for recycling. *The solution to pollution was dilution.*

But as Newark's population and industry continued to grow, the sheer volume of waste threatened to overwhelm the river's natural cleansing mechanisms. Soon, factories would begin dumping new kinds of waste into the river, waste that defied the natural cleansing process altogether. Manmade compounds such as PCBs and dioxins aren't just deadly, they're impervious to natural degradation. Water doesn't dissolve them. Microbes can't dismantle them. They cannot be deactivated. Once these compounds enter the environment they persist for decades, or longer.

The Passaic and its tributaries were no match for the quantity and character of this waste. In his 1975 book *Newark: The Nation's Unhealthiest City, 1832-1895*, historian Stuart Galishoff calls the mid-19th Century city "a vast cesspool of human and animal excrement and industrial wastes." Writing from the public health perspective, Galishoff paints an especially bleak picture. Even the upbeat Cunningham acknowledges that by 1870 Newark had an industrial-strength problem on its hands: "The sweet

Passaic had turned sour and filthy… The creeks and marshes were redolent with sewage. Water became nearly as precious as gold— and often as tainted and foul as the sewage-infested Passaic."

If ever a problem cried out for Newark's bold entrepreneurial spirit, for its inventiveness and business savvy, it was the problem of how to manage the city's effluent, which threatened public health as well as civic pride. Surely, the tinkerers of Newark would invent a solution that the city's government, with help from its financial industry, would then bankroll. "It was a time for Newark to keep the city from being ruined by its own progress," writes Cunningham.

But nothing happened.

Rather than confront the problem with characteristic verve and resolve, the city's political and business leaders largely ignored it. When city government did take action it was often tepid and too late. In 1850, Newark's governing council turned a deaf ear to calls for an underground sewer system and instead authorized the digging of an open drainage ditch as a way to ease the city's chronic waste and flooding problems. Editors at the *Newark Daily Mercury*, who were often and justifiably hard on Newark's elected officials, ridiculed the ditch as "another one of those halfway measures that has marked our city policy."

City policy had indeed shifted. As Newark's economy continued to industrialize, the levers of city government began to change hands. By 1850, the influence of Newark's aristocratic founding families was on the wane. Their dream of peaceful isolation had been eclipsed by the desire for robust commerce with the outside world and for the riches that commerce promised to bring. An 1860 census showed that nearly 75 percent of the city's workers held manufacturing jobs. Five years later, on the eve of the Civil War, Newark boasted the largest industrial-based economy in the U.S.

Gradually the composition of city government came to reflect this shift. More and more, candidates for public office were recruited from the world of business and industry. The result was a political-industrial partnership, which, though neither malevolent nor corrupt, produced a succession of civic leaders whose vision of government and government spending was limited in scope.

Newark wasn't alone in this regard. The entire state of New Jersey took a *laissez-faire* approach to business. Corporate taxes were low. Government regulation was minimal. The state's abundance of skilled labor and natural resources, such as rivers and creeks, helped turn Newark and other New Jersey cities into manufacturing centers. But workers and water weren't the only reasons. As early as 1882, in testimony before the New Jersey State Legislature, Ezra M. Hunt, then secretary of the state's Board of Health, suggested another explanation for why New Jersey was fast becoming a magnet for dirty business. Hunt used hats to make his point. The hat-making industry relied on volumes of toxic dyes and mercury-based chemicals to treat its cloth. "From year to year," Hunt told legislators, "there is an increasing tendency to locate such establishments in this State, in many cases, because New York and Philadelphia authorities will not allow them within city limits."

Unlike its neighbors, the pro-business state of New Jersey seemed to welcome all industry. Much of that industry set up shop in Newark. The city's credo, writes Stuart Galishoff, "that the most important function of government was the promotion of business, led to a narrowly circumscribed public policy that largely ignored health and social welfare concerns."

The stage was set for Diamond Alkali.

3 | DIOXIN

On the morning of June 2, 1983, the governor of New Jersey declared a state of emergency. Speaking at a press conference in his Trenton office, flanked by Newark mayor Kenneth Gibson, Essex County Executive Peter Shapiro and several state environmental officials, Governor Thomas Kean told reporters that the state's Department of Environmental Protection had detected disturbingly high levels of dioxin at the former Diamond Alkali chemical plant at 80 Lister Avenue in Newark's Ironbound neighborhood. With a three-page executive order, Governor Kean immediately shut down the Newark Farmer's Market, a major food distribution center about a block from the Diamond site. He stopped all train traffic around 80 Lister Avenue and expanded an already existing ban against eating fish or shellfish from the Passaic River. The governor stopped short of evacuation, but he offered temporary housing in Newark's YMCA to those residents who lived closest to the plant site, and he advised everyone to stay indoors during the cleanup operation.

The following morning, June 3rd, a dozen or so federal Environmental Protection Agency investigators, dressed in Hazmat gear, fanned out across the Ironbound. They searched the Diamond Alkali plant site, the adjacent banks of the Passaic River and the surrounding streets, homes, schools and businesses for any signs of stray dioxin. Press photos from the next few days show white-suited EPA workers literally vacuuming the streets of the Ironbound.

"It was like an invasion," recalls Nancy Zak. "All these guys in these moon suits were walking around. We're wearing our regular clothes. Nobody's telling us we should dress or do anything differently. It was a shocking day for people."

I met Nancy and her husband Arnold Cohen at Casa Vasca on a warm September evening in 2006. The popular Basque-style restaurant is at the corner of Elm and Prospect streets, about a half mile south of the river, and just a few blocks from the Ironbound Community Corporation (ICC), the nonprofit where Nancy works as a community organizer. Nancy and Arnold have been living in the Ironbound since 1972. That was the year Nancy arrived from Chicago to teach English at Independence High School, and Arnold started interning at a new ICC daycare program. Arnold now works as Policy Coordinator for the non-profit Housing Community Development Network of New Jersey. Nancy has been the ICC's Director of Community Organizing for more than 20 years. They have a teenage daughter, Beth.

In the weeks following Governor Kean's announcement, the men in the moon suits went house by house, street by street, factory by factory in the vicinity of 80 Lister Avenue. They collected dirt and weeds and street grit and the dust from vacuum cleaner bags and industrial air filters — 532 samples in total — and analyzed it all for the presence of dioxin. Meanwhile, local EPA officials tried to manage the community's growing fears and concerns. "Right now the problem is contained at the site," Dr. Richard Dewling, deputy administrator for the Federal Environmental Protection Agency Region 2, assured Newark's *Star-Ledger* two days into the sampling operation. "What we're doing off the site is strictly a precautionary measure to verify what we honestly feel: that this material has not migrated off the site."

It took all summer to finish collecting and analyzing the samples. When the EPA released the final results of its cleanup operation, investigators reported "massive" amounts of dioxin on Diamond's 80 Lister Avenue property, including a 60 parts per billion (ppb) level in the soil at the plant's south gate and a 51,000 ppb reading in the ground beneath an old storage tank. Dewling's prediction notwithstanding, EPA workers also recorded high levels of dioxin off-site — in an air duct at the abandoned waste treatment facility next door to the Diamond plant, and in dust from the vacuum cleaner bag of Carol De Francis, who lived nearby at 13 Esther Street. The off-site samples contained levels of dioxin ranging from zero to 15 ppb. At the time, the U.S. Centers for Disease Control considered concentrations above one ppb to present "an unacceptable risk to human health."

"[The dioxin] had attached itself to the trucks," says Nancy Zak. "Since that was an industrial area, the trucks were moving in and out, so they found dioxin in different parts of the neighborhood." In the community pool, for example, and in the "triangle," an area near the plant site where the oblique junctions of Raymond Boulevard, Ferry and Lockwood streets create an island of dirt shaped like a pizza slice.

By the summer of '83, dioxin summer, the residents of the Ironbound were already seasoned environmental warriors. The community had been battling more than half a dozen county and state proposals to build garbage, sewage and toxic waste incinerators in the neighborhood. It prevailed in most of the cases, though the state did eventually put a waste incineration plant in the Ironbound in 1991. When I ask Arnold and Nancy how they first found out about dioxin, Arnold insists on starting with the incinerator wars.

Back then, he explains, burning garbage was considered healthier than dumping it into the already bulging landfills in North Arlington and Staten Island. "We helped educate people that incineration had its own dangers," says Arnold. "We were part of a national movement in really changing the whole culture around what we do with our waste. I don't think the incinerator proponents expected the opposition they got."

The opposition came from the ICC's Committee Against Toxic Waste. Arnold was a founding member. The Committee mobilized the community around the incinerator proposals. It recruited experts. It organized informational meetings at the local Portuguese Sports Club. It held fundraisers like the benefit concert on the banks of the Passaic during which concert organizer Johnny Dirt, who owned a Bloomfield nightspot called the Dirt Club, was lowered into the river in a wet suit. ("The highlight," recalls Nancy, "was him coming out alive.") The Committee Against Toxic Waste bussed hundreds of Ironbound residents to public hearings so they could testify in opposition to incinerator plans, and joined forces with activist groups in nearby Bayonne, Jersey City and Elizabeth to form the Greater Newark Commission Against Toxic Wastes. "Those were the struggles that were happening in 1983 when the news about dioxin broke," says Arnold. "Everything we were doing was reactive. If we had had the time, we would have seen the signs [at Diamond Shamrock], because they were there. They actually had a fulltime dermatologist on site because people were getting chloracne[1]. That's how many problems they had — they had to have a doctor *on site*."

---

1 Chloracne is a disfiguring skin condition caused by exposure to high concentrations of dioxin.

The Diamond Alkali Company was incorporated in 1910 by Pittsburgh glass executives who saw potential in the soda ash business — soda ash being a key ingredient for making glass.

Diamond opened its first soda ash plant in Painesville, Ohio in 1911, the same year the New Jersey Legislature passed the state's first workman's compensation law. During the next 40 years, Diamond bought, built or operated half a dozen factories in five states, and expanded its product line to include caustic soda, which was used to make lye and detergents; calcium carbonate, which was used to smooth paints and make printer's ink dry faster; and chlorine, which was used for dry cleaning and for water purification.

The expansion of Diamond's product line was often driven by a creative use of waste. Diamond's calcium carbonate, for example, was a treated by-product of the company's soda ash production process. The chlorine that the company sold was created — then captured and packaged — during the manufacture of caustic soda.

Diamond developed new products too. In 1936, after experimenting with various production methods, Diamond began selling its own magnesium oxide. Within a few years, magnesium would become a key component in World War II bombs. As the war effort ramped up, the U.S. government tapped Diamond to operate one of its 12 wartime magnesium plants. Soon, Diamond was running three U.S. war factories.

When World War II ended, Diamond expanded its chemical business, mostly by acquiring existing chemical companies or divisions. By 1950, Diamond owned a dozen different plants that were making more than 100 different chemicals. And it kept right on growing. In 1950, the year my parents were married, Diamond acquired E.I. du Pont's de Nemours' chromic acid division. In 1953, Diamond purchased Belle Alkali, a company that made the starter chemicals for various resins and solvents. The acquisition of Belle Alkali put Diamond in the plastics business. In 1957,

Diamond acquired Diamond Black Leaf, a manufacturer of agricultural chemicals with factories in Texas, Alabama, Kentucky and Virginia.

Throughout the 1960s, Diamond bought and merged its way more deeply into chemicals and plastics, and eventually into coal and oil too. In 1967, Diamond merged with Shamrock Oil and Gas Company to become the Diamond Shamrock Corporation. During the next two decades, oil would surpass chemicals as the company priority. So much so that in 1986 Diamond Shamrock sold all its chemical holdings to the Los Angeles-based oil and gas conglomerate Occidental Petroleum, which is on the hook today for all Diamond Alkali's dioxin misdeeds. Ten years later, a more streamlined Diamond Shamrock merged with Ultramar Corporation, a Canadian oil concern, to become the Ultramar Diamond Shamrock Corporation.

Tracking Diamond Alkali's chemical history, with all its mergers, acquisitions and sales, is a little like opening a set of Russian nesting dolls. But for the purposes of this story, the most important chain of events began in March 1951. That's when Diamond Alkali made its initial foray into the emerging agro-chemical market by purchasing Kolker Chemical Works, a small manufacturer and distributor of herbicides and pesticides with a factory at 80 Lister Avenue in Newark.

There were still two vacant lots on Bathurst Avenue in North Arlington when I was growing up there. They were tiny wooded islands in a vast sea of houses, reminders of the once hardwood-forested hillside that plunged down to the Passaic. These pygmy forests, less than an acre in total, had managed somehow to escape the builders. Looking back, they remind me of the woodlots I saw in northwestern Ohio when I was a graduate student there at Bowling Green State University in the late-1970s. The local farmers grew corn mostly and they planted it everywhere in the flat windswept landscape. The only corn-free zones were the farmhouse perimeters and the slim skirt of vegetation that clung to the banks of the local Maumee River. But for some reason, the farmers would leave square patches of woodland standing here and there in the cornfields. A botany professor took us to one of the nearby woodlots in search of spring wildflowers. I saw my first trillium and Dutchman's breeches on that field trip.

The evolutionary biologists on campus used the woodlots as laboratories to study a phenomenon they called "island biogeography." Sea islands — the Galapagos, for example — are effective barriers to migration. As such, they are also potential breeding grounds for new species. Unless an island-dwelling species can fly or swim long distances, or somehow stowaway, its members are destined to become physically, and then genetically isolated from any off-island kin. Over time, the separation can lead to the evolution of new varieties and even new species. Bowling Green researchers reasoned that a vole stranded in an Ohio woodlot wasn't all that different from an iguana marooned on some equatorial atoll. I'm not sure whether they ever discovered a new species of vole or squirrel in one of those Ohio woodlots, but the idea always intrigued me.

The Bathurst woodlots were much smaller than the ones in northwestern Ohio, and they weren't nearly as pretty or as populated. They were home to a scruffy mix of second-growth maples and locust trees and shrubs and weedy grasses — and the occasional midden of empty beer cans and pint bottles of Southern Comfort. But to us, they were wild places. We called them "the lots," and declared them our own. In the warm months, we played hide and seek and designed obstacle courses. Winter was for sledding. In the fall, we built makeshift bleachers between the trees where we'd sit and root for the older boys in the neighborhood who raced homemade go-carts along the dirt footpaths. Both lots were gone by the time I graduated from high school in 1970, lost to more single-family homes. I'm surprised they hung on that long.

People like me who grew up in the cramped northeastern part of New Jersey reacted to the region's relentless development in different ways. Some moved "farther out," mostly west or north to leafier, more open suburbs like Florham Park or Chester or Ringwood. (My father's parents did the same thing back in the 1930s when they abandoned Jersey City for the then greener, emptier pastures of Belleville.) Some, like me, moved farther still — to Seattle or California or Florida. Some stayed put and coped with the crowding by escaping to the Jersey Shore for two weeks every summer. A few, like Michael Gordon, decided to stay and fight.

Gordon greets me in his law firm's reception area and leads me down the hall to a large conference room. He apologizes for the boxes stacked against one wall. The firm has just moved from West Orange to Springfield, New Jersey. The new offices occupy a three-story, red brick behemoth on Morris Avenue, Springfield's main drag.

In Union County, as in much of northeastern New Jersey, towns run pretty much back-to-back. There are a few undeveloped pockets, most notably the 2,000-acre Watchung Reservation that separates Mountainside from New Providence and Berkeley Heights. But driving this part of Jersey isn't like driving through the west or the Midwest or western Pennsylvania. There's no country land, no cornfields or mountains or prairies to buffer one town from the next, to set the towns apart. You just wind through suburbs, and then a downtown core, and then suburbs and another downtown, and so on and on. But if you pay attention, there is a discernible transition zone. It's a space, sometimes just a block or two long, where the outskirts of two towns meet. The place where you might find an A&P or an office building, or a gas station, or a diner, or a carpet store, or a mom-and-pop factory that makes custom sliding glass doors — or maybe all of them anchoring a wide intersection or elbowed together in a little stretch mall or maybe just scattered on either side of the same house-free thoroughfare that was just one town's main commercial street and is about to become another's.

Michael Gordon's office occupies such a zone. It's a pleasant, leafy interlude between Millburn and Springfield where a handful of new brick office buildings face each other across a four-lane roadway. Gordon's new headquarters has a spacious asphalt parking lot and two large signs on its north and west-facing facades that announce *Gordon & Gordon PC Attorneys at Law* in bold, blue and white serif type. With his partner and brother Harrison, Michael Gordon has been successfully litigating environmental health cases for more than 25 years. If the new office is any indication, environmental law — in New Jersey, at least — must be a pretty good business.

Gordon takes the chair at the head of the long wooden conference table and waits patiently while I unpack notebook, pens and tape recorder. Even seated, Michael Gordon looms large. He stands well over six feet tall. His dark hair is graying at the temples and thinning on top. He wears round tortoise-shell glasses. He doesn't smile much. For an award-winning attorney who has represented plaintiffs in some famous cases (Pan Am Flight 103, for one), and served as a consultant on the old NBC political drama *The West Wing*, Gordon is modest and understated, a man of few words, an anti-zealot.

Gordon has served as lead or co-counsel on almost every high profile environmental case in New Jersey. He has represented victims of radon in Montclair, lead paint contamination in 22 New Jersey cities and dioxin in the Ironbound. It was the dioxin case that launched his law firm and helped forge his reputation as one of the state's preeminent environmental attorneys. "My relationship with the Ironbound community changed my life," he says.

When Gordon graduated from Rutgers University School of Law in 1979 he weighed several job offers before taking the lowest-paying one—with the New Jersey Department of Environmental Protection. Environmental law intrigued Gordon, in large part because in 1979 the field was brand new. Rutgers' law school didn't even offer a course in environmental law until 1977, which was Gordon's first year. But like me, Gordon had grown up in northern New Jersey in the 1950s and '60s watching the last of the area's open space vanish. He was born in Newark, where his mother's family ran a chicken market in the Ironbound. He was raised in West Orange. The state built Route 280 right across the street from his childhood home. This up-close view of development made a deep impression on Gordon. It's a "pattern you see playing over and over," he says.

Gordon worked for the New Jersey DEP in Trenton for three years. During that time, he helped draft New Jersey's first Environmental Cleanup Responsibility Act, or ECRA. The law, which now goes by ISRA, the Industrial Site Recovery Act, basically requires industry to clean up after itself. Factory owners or operators now have to go through a kind of environmental checkout before they can sell, transfer or shut down their operations. If state inspectors find any environmental contamination, owners have to clean it up before they can unload the property. Prior to ISRA, says Gordon, "there were so many transfers where the new owner was supposedly surprised that he had bought a contaminated piece of property." According to Gordon, tying environmental cleanup to real estate transactions, "has accomplished thousands of cleanups" that would otherwise never have happened. Too bad ISRA wasn't around in 1971 when Diamond Shamrock sold 80 Lister Avenue.

New Jersey's original ECRA, the law that Gordon helped write, was the first and toughest of its kind in the country. ISRA is the milder 2.0

version. "Our laws were substantially weakened in the early 1990's in order to promote city revitalization," explains Gordon. Even so, the state of New Jersey has some of the strongest environmental regulations in the country, which seems fitting, if ironic. Why shouldn't the state that was at the forefront of environmental destruction be in the vanguard of environmental protection? If only there were a little more environment to protect, especially in my old stomping ground. The great marsh that once enveloped Newark, for example, is gone for good.

Gordon left the DEP for private practice in 1982. Working in a bureaucracy didn't suit him. Besides, after sitting through countless public hearings as a DEP rep and being accosted afterwards by citizens desperate to find a good lawyer to argue their cases, he saw a market niche. "I decided to represent those groups who were being confronted with pollution," he says.

A few months later, Gordon found himself at one of those very same public hearings, this one in Montclair. He wasn't there on official business. He had just stopped in to say hello to some of his old DEP colleagues. At the meeting, one of those colleagues introduced Gordon to members of the Ironbound's Committee Against Toxic Waste who — surprise-surprise — were looking for a lawyer. "So I said yeah, come by and see me," Gordon recalls. "The next day there were three of them in my office saying, 'we got this problem and this problem and this problem,' and within the span of a year (1982-83) the Thomas Street warehouse caught fire, with 17,000 drums containing mostly chemical waste next to a grade school; At Sea Incineration wanted to set up a loading point in the Ironbound to take hazardous waste out to sea and burn it; and a hazardous waste facility blew up and wanted to get a permit to rebuild even though they could no longer satisfy the environmental criteria. So I was their lawyer, pro bono, and we were successful on those three cases."

Then came Governor Kean's dioxin press conference on June 2, 1983, the day Michael Gordon got swept up by a wave he's still riding. (Gordon is currently Special Counsel to the State of New Jersey on the Passaic River dioxin matter.) "Those other cases were prominent," says Gordon, referring to the Thomas Street warehouse explosion, At Sea Incineration, etc. "But when the discovery of dioxin occurred, I mean, that was a huge, huge issue." The state-of-emergency orders, the shuttering of the Newark Farmer's Market, the guys in the Hazmat gear, the swarm of media. Litigating the dioxin case would be a whole different ballgame, a daunting challenge for Gordon and his fledgling firm of one. "I had no secretary and no funding," he says. "I had to determine a way to get up and running."

As it approaches Newark, the Passaic River loops sharply north and then just as sharply south again, as if the river is recoiling from the city. The Passaic's S-shaped course around Newark creates a unique polyp of Ironbound land that is bordered on three sides by water. Diamond Alkali's plant at 80 Lister Avenue sat at the edge of the polyp, on 3.4 acres along the western bank of the Passaic. Diamond shared its bulbous peninsula with three other chemical manufacturing companies. To the east there was Sergeant Chemical, which Diamond eventually bought. To the south and west was the paint giant Sherwin-Williams. In the southwest corner was Duralac, Inc., a small producer of varnish, lacquers and enamel coatings. All the land occupied by these four chemical companies lies within the Passaic River flood zone. Sherwin-Williams is the only one of the four still operating, albeit in a different locale, several blocks south on Raymond Boulevard. Benjamin Moore Paints occupies the old Sherwin-Williams space now.

From the water, this stretch of Ironbound riverfront looks unquestionably industrial. But it is well kept and still active compared to some of the abandoned, blighted zones to its south. The Benjamin Moore factory, with its fresh coat of white paint and corrugated tin roof, hugs the riverbank just behind a low concrete bulkhead, which has been whitewashed to match the building. A hurricane fence trimmed in barbed wire runs along the top of the bulkhead. Next door, on the upstream side, is the Diamond superfund site. It certainly isn't pretty, but again it's neat, and in this neighborhood that counts for something. The concrete bulkhead here is pale gray, higher than the Benjamin Moore wall, and newer. From the river, you look up at a gray gravel surface that slopes back and upward from the wall like a wide, groomed ramp to nowhere. Stark white planters — two dozen or more — have been placed around the mound. Each contains a small evergreen tree, forlorn perhaps for being stranded alone in its white concrete cell. This attempt at landscaping reminds me of one of Governor Kean's warnings in the summer of '83. He told Ironbound residents not to eat anything from their gardens for fear that dioxin in the soil may have poisoned homegrown fruits and vegetables. There's no barbed wire on the hurricane fence at the Diamond site. What's buried here keeps would-be thieves and vandals away.

The view from the Lister Avenue side is much starker: weed strewn lots and crumbling factories, shattered windows and higher hurricane fences, padlocked gates, thickets of razor wire and large signs warning KEEP OUT or BEWARE OF DOG or NO DUMPING: YOU ARE BEING VIDEOTAPED. The old Duralac factory is still standing. It's a two-story rectangular box of beige brick with D-U-R-A-L-A-C in sky blue letters. The "U" has begun to pull away from the brick facing. Someone has spray painted the number 84, in black, on the flaking white door in the corner of the building that faces Lister Avenue. A graffiti address.

From here, you can see the tall stack of Essex County's massive waste incinerator, the one that Arnold Cohen and Michael Gordon fought so long and hard against. The incinerator sits near the riverbank at the far, far eastern edge of the Ironbound, in a desolate field of dirt mounds at the very end of Blanchard Street.

"This is the very last street in Newark," says Carol Johnston, "and this is where the incinerator is." Johnston is a 60-something Catholic nun who is the Director of Special Projects for the Ironbound Community Corporation. She is a whip-smart, fast-talking Newark native, who grew up in a blue-collar family in the city's Vailsburg section. The major focus of her work with the ICC is an ambitious riverfront development project. Johnston is trying to sell city, county and state officials, plus a handful of private landowners on a plan to convert 22 acres of Ironbound land into a waterfront park. We are on a driving tour of the properties in play.

Johnston zooms around these peninsular Ironbound streets in her white Nissan Sentra, crisscrossing Raymond Boulevard and stopping now and then to let me hop out and snap a picture. As she drives she delivers a non-stop commentary on the intricacies and intrigues of the waterfront development plan ("The city agreed to make the old Newark Fire Training Center and heliport open space and then reneged in favor of development."); on the gentrification of the Ironbound, in spite of its superfund history ("The community has been transformed into a much more volatile place with staggering house values and high rents."); and on the legacy of Newark's formerly corrupt political machine ("There's no planning in Newark; what there are are deals."). I also learn that I can buy drugs at the Sabrett hot dog stand on the corner of Fleming Avenue and Ferry Street; that Johnston's grandparents were married in St. Aloysius Catholic Church on the corner of Fleming Avenue and Freeman Street; and that the 15-acre riverfront lot stacked full of empty shipping containers is a stark reminder of America's negative trade balance. "We have nothing to send the world in these things," says Johnston, gesturing in the direction of the containers. "So they get piled up all over." When I ask about the steady trickle of men and women I notice wandering down Blanchard Street, a bleak wasteland of a road that dead ends at the river, Johnston sighs. "You only see people walking up and down this street for one reason," she says: "The methadone clinic. It's right down the street from the waste incinerator, which says what we think about both of them."

This swatch of Ironbound framed by Raymond Boulevard, the river, Chapel Street to the west and Blanchard on the east is a crush of homes and heavy industry. The factories, arrayed along Lister Avenue, command the riverfront. Behind them, modest homes line Cornelia, Esther, Joseph, Albert and Euclid streets. Industrial manufacturing—mostly of farm chemicals—has been going on at 80 Lister Avenue since Alfred and Joseph Lister bought the property in 1850. The brothers built the Lister Agricultural Chemical Works factory where they ground cattle bones into fertilizer. Kolker Chemical Works acquired the Lister Avenue property in the early 1940s. Kolker made farm chemicals too. The pesticides and herbicides it produced on the site were not so benign.

Kolker manufactured DDT and the phenoxy herbicides 2,4,5-T (2,4,5-trichlorophenoxyacetic acid) and 2,4-D (2,4-dichlorophenoxyacetic acid). DDT is a pesticide that killed insects by attacking their nervous systems. As a kid in the 1950s, I remember running behind the small truck that would drive through our neighborhood on summer evenings spewing a white industrial-smelling fog from a noisy little machine mounted in its bed. The billows of DDT fumigant it sprayed were meant to control local mosquito populations. We called the truck the "skeeter beater."

Diamond stopped making DDT in the late 1950s, more than 20 years before the Environmental Protection Agency banned use of the pesticide. From the late 1950s until 1969 when Diamond closed the Lister Avenue plant, phenoxy herbicides were the only products manufactured there. Phenoxy herbicides were the ingredients in Agent Orange, which was a 1:1 cocktail of 2,4,5 T and 2,4 D.

Phenoxy herbicides were weed killers. They dispatched their botanical victims by mimicking a plant growth hormone. When sprayed on broad-leaf plants — they only work on broad-leaf and woody plants — phenoxy herbicides induced a rapid, distorted and uncontrollable growth. Like a cancer, this wild, sustained burst eventually destroyed the plant. The speed with which some phenoxy herbicides could kill vegetation was discovered by Professor E.J. Krauss, chair of the University of Chicago's botany department during the Second World War. Believing the compounds might have wartime applications, like destroying enemy crops, Krauss shared his findings with the U.S. War Department.

The American military never did use phenoxy herbicides during WW II. But after the war, U.S. chemical companies quickly saw their peacetime potential as commercial weed killers for fields, roadsides, lawns and gardens. So began America's love affair with synthetic herbicides.

Michael Gordon knew he couldn't tackle the dioxin case by himself. The discovery process alone — getting access to all the relevant information and then reviewing it — was daunting. In addition, he'd have to recruit (and pay) a team of credible experts who could demystify dioxin. So little was known about dioxin at the time that New Jersey initially put off investigating potentially contaminated sites because there was no chemical assay sensitive enough to detect dioxin in the minute quantities at which it posed a health risk. Gordon didn't have the money, the staff or the experience for such a complicated undertaking. But he knew someone who could help.

Melvin Belli was the flamboyant California "King of the Torts." Best known as the lawyer to the stars, both famous and infamous, his client list featured Errol Flynn, Lana Turner, Mae West, Chuck Berry, The Rolling Stones, the Bakkers (Jim and Tammy Faye), Muhammad Ali, George Forman, Evil Kneivel, Sirhan Sirhan and Jack Ruby. During his long career, Belli had also been a pioneer in consumer rights law. Although they never met, Gordon knew of Belli because both had worked on the Montclair, New Jersey radon case. Belli was one of the first attorneys to use visual aids, or "demonstrative evidence" in court. One of his earliest trials involved representing Chester Bryant, a San Francisco cable car worker who'd been seriously hurt when the metal grip connecting his cable car to its power line snapped and slammed into his stomach. At trial, Belli hauled an actual cable car gearbox into the courtroom along with a detailed model—as big as a king-sized bed—of the intersection where Bryant was injured. The jury found for his client.

"He was already beyond the peak of his career," says Gordon about Belli, who was 76 in 1983. "But he was one of the most prominent and active trial lawyers in the country. And he really knew how to go through process." Belli also had a Rolodex of expert witnesses and he was at home in the spotlight. Gordon really needed somebody like that. So he picked up the phone. "I called and said I have an interesting case. Would you want to team up?" Gordon recalls. "He said, 'why don't you come out?' So I used all my money to fly to San Francisco."

Gordon arrived on a Friday. He didn't get to sit down with Belli until Sunday afternoon. He made his pitch while they watched a San Francisco 49-ers game in the living room of Belli's home, which overlooked the Golden Gate Bridge. The 49-ers lost to the Redskins that afternoon. But Gordon left San Francisco with a silent partner. Belli agreed to help locate and finance the scientific and health experts who could assess the cause, extent and impact of the dioxin contamination. Belli died in 1996. Today, there are lots of large law firms with the resources and the appetite to go one-on-one with corporate America. That legal infrastructure didn't exist in 1983. Dioxin helped create it. The Ironbound case, says Gordon, "was the first time, at least in New Jersey, that someone brought together community activists, legal talent and scientific talent that could match the industry."

In August 1983, Gordon filed two dioxin-related lawsuits on behalf of the Ironbound community. He sued his old employer, the state's Department of Environmental Protection, as a way to compel the agency to modernize and organize so that it could properly direct Diamond's cleanup efforts. He sued Diamond for damages, and for a peek into the company's files. "This was one of the earliest hazardous waste cases in New Jersey," says Gordon. "It was hard to get information then."

Fortunately for Gordon, he wasn't the only one eager to delve into the chemical company archives. A group of Viet Nam veterans frustrated by government inaction had just filed suit against Diamond Alkali, Dow Chemical, Monsanto and other manufacturers of Agent Orange. The plaintiffs — more than two million veterans — were after compensation for health problems they blamed on wartime exposure to the herbicide. Their 1982 class action lawsuit forced Diamond and its co-defendants to hand over reports, memos, letters, invoices, manufacturing diagrams, every scrap of paperwork that might shed light on the claims. Conveniently for Gordon, the presiding judge in the Agent Orange suit, Jack Weinstein of New York's Federal District Court, had ordered that all company files be stored in the same place — the law offices of Diamond's attorneys Cadwalader, Wickersham & Taft, which overlooked the Hudson River at 200 Liberty Street in lower Manhattan.

Gordon's discovery process and the decontamination of the Ironbound proceeded in tandem over the next few years. While teams of private contractors dismantled and sealed the Diamond plant site, and Ironbound activists agitated for more and better public health studies, Gordon spent his days in the Cadwalader offices sifting through box after box of yellowed paperwork. He was looking for a memo or a letter or a report, some written evidence that Diamond had been aware of its dioxin problem and yet had done nothing to address it. It was a tedious job. But one afternoon in 1986 Gordon found the smoking gun. It was a file of correspondence and related internal memos between Diamond employee D.J. Porter and officials at the C.H. Boehringer Sohn chemical company of Germany. The exchanges left no doubt that Diamond managers knew dioxin was contaminating their 2,4,5-T manufacturing process; that they knew dioxin was the cause of the chloracne outbreak among workers at the Newark plant; that they knew how to eliminate the dioxin from the 2,4,5-T reaction; and that they knew all of this as early as 1957. "I went back and read the file a few times," says Gordon. "And I remember saying, okay, that's it. I'm done."

Diamond Alkali plant workers used a three-step process to make their batches of 2,4,5-T, one of the two ingredients in Agent Orange. Workers began with a manmade starter compound called 1,2,4,5-tetrachlorobenzene. This odorless chemical came in several forms — colorless crystal, flaky white powder, or solid chunk — and it too was toxic. Workers mixed the tetrachlorobenzene with sodium hydroxide and methanol to produce a sodium salt called 2,4,5-trichlorophenol. Trichlorophenol, or TCP, was then combined with sulfuric acid to yield the desired end product: 2,4,5-T. All the reactions were carried out at high heat and under great pressure in an autoclave, a kind of laboratory-grade pressure cooker. The most delicate phase of 2,4,5-T production was the very first step, the conversion of tetrachlorobenzene to trichlorophenol. That's where the dioxin was formed. It was one waste product that Diamond would never find a use, or a market for.

Though dioxin was first synthesized in a German laboratory in 1872, it really is a 20th Century phenomenon, a nasty little legacy from the age of chlorine. As early as the 1920s, organic chemists began stitching chlorine atoms to hydrocarbon molecules. In so doing, they created whole new classes of chemicals, from oily substances like PCBs to powerful insecticides like DDT to unintended poisons like dioxin.

The most dangerous dioxin, TCDD, was the one being generated in massive quantities by workers at 80 Lister Avenue. TCDD's very long name, 2,3,7,8-tetrachlorodibenzo-para-dioxin, is actually a helpful short hand for its molecular structure. Like other members of the dioxin clan, TCDD has two benzene rings (hence, the dibenzo in its name) coupled by two oxygen atoms (dioxin). What makes TCDD unique is the number and arrangement of its chlorines. TCDD has four chlorine atoms (tetracholoro) located on the carbons in ring positions 2, 3, 7 and 8. This particular arrangement gives TCDD a perfect symmetry. The 2,3 and 7,8 positions occur at the front and rear corners of the molecule. TCDD's chlorines are like the headlights and tail-lights on a car. (The "para" in 2,3,7,8-tetrachlorodibenzo-para-dioxin's name refers to the position of the oxygen atoms.)

Temperature, as it turned out, was the key to preventing TCDD contamination of the Agent Orange batches. Above 320 degrees Fahrenheit, trichlorophenol becomes unstable. The chemical bonds that hold its chlorine and carbon and oxygen atoms together come undone. The compound unravels.

Carbon is a promiscuous element; it likes to bond. The breakup of trichlorophenol presents carbon with bonding opportunities galore. New molecular matchups lead to new compounds—2,3,7,8-tetrachlorodibenzo-para-dioxin is one of them. The hotter the reaction, the more TCDD is formed. Trichlorophenol should be simmered, not boiled.

Based on the paperwork Michael Gordon unearthed at the Cadwalader offices that day, it became clear that Boehringer knew all about unwanted dioxin formation. Workers at its chemical plants began developing chloracne in the late 1940s. Boehringer chemists correctly identified the culprit as a contaminant in the trichlorophenol reaction. By lowering the heat in the reaction vessel and tweaking a few other parts of the process, they eventually eliminated the dioxin and chloracne problems. Boehringer was eager to share its experience and know-how with Diamond and with the other American chemical manufacturers who were working with trichlorophenol.

In a September 18, 1957 memo to John Burton, the Newark plant manager at the time, Diamond's D.J. Porter relays information he received from a Boehringer official who had just visited Diamond's Painesville, Ohio factory:

"Dr. Kudszus of Boehringer was here last week," writes Porter. "He pointed out that a compound identified as an extremely toxic material and probably the cause of some of the chloracne difficulties…is not formed if the trichlorophenol autoclave temperature is kept below 155 degrees C and if the dried TCP-caustic crude mixture is not allowed to exceed 120 degrees C… In view of our problems with chloracne in our trichlorophenol plant, it would certainly seem in order to check out these observations, if possible, in trial plant runs." In anticipation of Burton's concern about lower reaction temperatures slowing down production, Porter goes on to propose an autoclave redesign that "might be able to affect an actual increase in our production capacity…."

Diamond envoys made several trips to Boehringer's Frankfurt factory in the late 1950's. Thornton F. Holder visited Boehringer in the fall of 1959. His September 30th memo, also to John Burton with copies to D.J. Porter and others, reveals how alarmed Boehringer was over the potency of dioxin. "[Dr. Kudszus] says the dioxime [sic] is so active as to be a chemical warfare agent," writes Holder.

Boehringer's fear of dioxin's toxicity drove company officials to modify not only the manufacturing process but the workplace as well. The changes worked. "Boehringer has had no new cases of chloracne in 3-5 years," writes Holder. "They did this by discovering the compound which is the bad actor and eliminating it as a by-product ... They then cleaned up their factory in a major fashion before reopening — it was closed a year...In addition, even though Boehringer is sure it has no more dioxime [sic] in its process, it ventilates its plant with great care, issues clean work clothes every day and continuously checks its trichlorophenol on rabbit ears... Finally, after Boehringer did all of the above, they tore out plaster walls, floors and similar parts of buildings until no remaining structure gave positive rabbit ear test. This done by carefully protected workmen." Holder concludes his memo to John Burton with a bit of cheek: "Better try a few rabbit ears on your joint if you are still having trouble."

Although Diamond took Burton's suggestion and performed some rabbit ear tests, they never lowered the temperature of the 2,4,5-T reaction. The company did eventually hire an on-site dermatologist so that workers wouldn't lose so much time traveling back and forth to doctors' appointments. But when it came to eliminating the cause of the chloracne, Diamond ignored Boehringer's advice along with suggestions proposed by its own managers in the flurry of internal memoranda. Even after a February 1960 autoclave explosion that killed one Diamond worker and forced the company to rebuild its Newark plant—autoclave temperatures soared off the scale at 392 degrees Fahrenheit—Diamond didn't change a thing.

"They thought they were dealing with it effectively," says Michael Gordon, about Diamond's non-response. "That it wasn't a big deal. That occupational exposures were a price of manufacturing." In the run-up to trial, Gordon deposed several Diamond officials, including Dr. R.W. McBurney, the company's medical officer, and Ray Guidi, the man who succeeded John Burton as manager of the 80 Lister Avenue plant. "They never really came up with a viable theory for why they didn't implement the changes," says Gordon, who believes nonetheless that Diamond's decision not to act was driven by bottom line concerns. New Jersey Appellate Division Judge David S. Baime put it more bluntly in his 1992 opinion in the *Diamond Shamrock Chemicals Company vs. Aetna Casualty & Surety Company* case. (Diamond sued Aetna, its insurer, after Aetna refused to indemnify claims made against Diamond.) Writing for the majority, Judge Baime noted: "The only conclusion to be drawn is that Diamond's management was wholly indifferent to the consequences flowing from its decision. Profits came first."

In 1967, a decade after it was first warned about dioxin, Diamond finally took some action. By then the company was one of the largest Agent Orange manufacturers in the U.S. The 80 Lister Avenue factory turned out 16 million pounds of tricholorphenol between 1952 and 1968. With its Agent Orange production at full tilt for the war effort, Diamond upgraded the ventilation system in its TCP building. Noxious vapors that had been wafting directly into the air over the Ironbound for years were, for the first time, funneled through a carbon tower that was designed to trap dioxin.

Michael Gordon credits the addition of a carbon tower at 80 Lister Avenue to "competitive pressure." Industry insiders knew about the high levels of dioxin in Diamond's 2,4,5-T product. A few Diamond customers had actually developed chloracne after handling the company's 2,4,5-T and 2,4-D herbicides. Reports like that were bad for business—*everybody's* business. In 1965, the Dow Chemical Company tried to prod Diamond and other U.S. chemical manufacturers into setting an industry standard for dioxin levels in 2,4,5-T. Like Boehringer, Dow had reduced dioxin formation in its 2,4,5-T process and taken steps to protect its workers. Dow proposed an industry standard of one part per million (ppm) for dioxin, a level far below the concentrations of dioxin that were still typical in Diamond's 2,4,5-T. Some batches showed levels as high as 26ppm. "Dow was very concerned that there was so much dioxin in Diamond's product," says Gordon. "The Agent Orange program was really booming, and they were worried that Diamond's product would undermine the business."

Diamond didn't bow to the pressure, essentially calling Dow's bluff. Dow never squealed on its competitor, and Diamond never bought into Dow's industry-wide dioxin standard. Nor did Diamond ever try to remove dioxin from the trichlorophenol reaction, Boehringer's successful front-end fix. Chloracne continued to plague the company's workers right up until they turned out the last batch of TCP in June of 1969. That's when Diamond shut down Agent Orange production. Diamond closed the Newark plant altogether in August of that year. Two years later, in March 1971, Diamond sold the property to Chemicaland Corporation, which made benzyl alcohol, a multipurpose organic solvent used by ink, paint and lacquer manufacturers. The sale marked the end of the dioxin production at 80 Lister Avenue.

It took a while, 14 years to be exact, but dioxin came back to haunt Diamond. The company, then the Diamond Shamrock Corporation, spent most of the 1980s in litigation over dioxin. There was Gordon's Ironbound case and the Agent Orange lawsuit brought by Viet Nam vets. In 1989, Diamond sued its own insurance company, Aetna, over Aetna's refusal to cover the cost of the 80 Lister Avenue cleanup and the $23 million settlement with Vietnam veterans. The *Diamond vs. Aetna* case was eventually appealed to the New Jersey Supreme Court, which found for Aetna in April 1992. In the deluge of documents, depositions and media stories unleashed by all the litigation, the truth about operations at Diamond Alkali came tumbling out. It became painfully clear that at 80 Lister Avenue the only thing more careless than the chemistry was the housekeeping.

Before 1956, all the plant's chemical waste went — untreated — into the Passaic River. Several spills sent as much as 30,000 gallons of acid *a day* into the water. The Passaic Valley Sewerage Commissioners installed an industrial sewer line in the area in 1956. After Diamond was hooked up to the line, it continued to foul the river. In fact, before it stopped producing DDT in 1959, Diamond pumped so much DDT wastewater into the Passaic that a "mountain" of the pesticide began rising in the middle of the river. Fearful that passing boaters would report the strange mound, plant managers sent workers wading into the river during low tides to "chop up" the DDT.

Plant practices didn't improve much after Diamond stopped making DDT. In testimony for the Aetna case, Diamond plant worker Chester Myko called the floor in the TCP building the "dirtiest place in the entire plant." The amount of 2,4,5-T and 2,4-D that was regularly slopped onto the plant floor made the surface so slick that walking across it was treacherous. Workers hosed down the floor with sulfuric acid every week or so, directing the wastewater into open trenches, which took it outside the building and into the Passaic.

Arthur Scureman was a Kolker Chemical Works employee who stayed on after Diamond purchased Kolker. Scureman worked at the plant until it closed in 1969. He described numerous leaks in the autoclave room itself and in the pipes that carried caustic waste out of the autoclave room — and eventually into the Passaic.

Aldo Andreini worked for Diamond from 1959 to 1969. His job was cleaning the 10,000-gallon tanks that were used to store the Agent Orange ingredients 2,4,5-T and 2,4-D. Once or twice a month, Andreini shoveled sediment from the storage tanks into metal drums. The liquid and solid waste that got spilled onto the ground during the transfer would be washed away—into the Passaic. Andreini also cleaned the railroad cars that shipped the finished Agent Orange product. He would rinse the inside of the cars with water, which flowed out of the cars and onto the tracks and along the ground—and eventually into the Passaic.

Harry Heist was employed at Diamond from 1966 to 1969. "There were spills all the time," he told the *Star-Ledger* in June 1983. "The reactors would run away, boil over and stuff would flow down the sides of the tanks and troughs on the floor that led to the river. The stuff was all over the place."

In the words of its own workers, Diamond's waste disposal strategy amounted to "dumping everything" into the Passaic River.

Chloracne was the visible price that Diamond factory workers paid for this carelessness. In 1955, Diamond sent its corporate medical officer, Dr. William York, out to the 80 Lister Avenue plant to investigate what company files characterized as a "major outbreak of chloracne." York called the chloracne a "serious… very disfiguring social disability." He couldn't determine the exact cause of the chloracne, but he suspected some chemical agent and he made several common sense recommendations about how to avoid worker exposure. These included keeping the work area cleaner and designing some kind of specialized spout with a suction device that workers could use whenever they poured toxic liquids and powders from one container to another. York also cautioned Diamond officials that chloracne might just be the tip of the iceberg. Dioxin, he said, had the potential to incite other, less visible but more systemic health problems among Diamond workers. The doctor's recommendations and warnings went unheeded.

York, of course, was right about dioxin's potential for widespread toxicity. In the more than 50 years since he first warned Diamond officials that the unknown chemical contaminant in their Newark factory could harm workers in other more profound ways, epidemiologists have found that people exposed to dioxin — as a result of industrial accidents in Germany, the Netherlands and the United States — are far more likely to die from cancer. Researchers working with laboratory animals have determined that exposure to tiny amounts of dioxin causes cancer, birth defects and permanent damage to the immune system. Animal studies show that no matter how the dioxin is administered, whether it is eaten, inhaled, injected or dabbed onto the skin, exposure produces every type of cancer, in every strain and every gender of every animal tested — every rat, mouse and hamster. Exposed animals developed cancers in their lymph and respiratory systems, in the liver, skin, thyroid, adrenal cortex, nose and tongue. Science has now confirmed what Dr. William York and others suspected in 1955 — dioxin is a potent and indiscriminate poison.

As Michael Gordon prepared the Ironbound case for trial, the IT Corporation, hired by Diamond and supervised by the state, took the 80 Lister Avenue plant apart. IT workers broke down and gathered up all the equipment that was used to make Agent Orange. They packed up every square inch of flooring and wall board, every pipe and valve, every beam, brick, gate, fencepost, lock, chain, window, door, shelf, closet, light fixture, sink, toilet, desk, filing cabinet, phone, wastepaper basket, clipboard, pencil. Everything. They laid the bits and pieces of Diamond Alkali to rest, without ceremony, right there at 80 Lister Avenue in the six-acre, clay-sealed, membrane-capped, gravel-topped grave with the white planters on top. Federal Environmental Protection Agency officials proposed cremating Diamond Alkali on site. Ironbound residents soundly rejected the EPA's "on-site incineration" plan as unsafe. Instead, the remains of Diamond Alkali were sealed inside 032 shipping containers along with the neighborhood's toxic dirt and the dust from Carol De Francis's vacuum cleaner bag.

The Ironbound Committee Against Toxic Waste staged meetings and rallies throughout the cleanup operation. The community protested the contamination. They hounded local officials to monitor the cleanup more closely. They demanded more health testing and long-term health studies. It was a confusing, exhausting, emotional time in the Ironbound. When reporters came looking for neighborhood reactions, some longtime residents pooh-poohed the dioxin scare. "I've been here 21 years and I'm still alive," said Marilyn Borella, who lived on Esther Street, just a few blocks from the plant. Her neighbor, Estelle Trechel, who was born and raised on Esther Street in a house that was barely a block from Diamond Alkali, agreed. "If there was anything around that would have made us sick," said Esther, "we would have caught it six times over."

Nancy Zak was skeptical too, at first. The guys in the moon suits seemed a little over the top, almost comical. But then, says Nancy, "we had this series of marches on Ferry Street. A former student of mine from Independence High School came to one of those marches. Diane. I can't remember her last name. She was a young mother. She had two kids born with birth defects and she herself had developed serious liver and eye problems. She was losing her vision. She had moved out of the Ironbound. We just started talking about why she was at the march and how she'd heard about it and she thought that what had happened to her had something to do with the dioxin. She had played on the piles of dirt in that neighborhood. The kids would run up and down on those hills where the trucks had tracked all this stuff. I was always reading about how chemicals can hurt you and suddenly there was Diane. For me, that was really the beginning of understanding what chemicals could do to people's health."

Complex manmade chemicals such as dioxin are especially dangerous. Toxicologists often refer to such compounds as xenobiotic, literally "foreign to life." Xenobiotics are substances, usually chemicals, that don't exist naturally in the mammalian body. Not all xenobiotics are bad. Antibiotics fall under the rubric. But all xenobiotics are, by their very nature, exotic and in evolutionary terms quite new. This unfamiliar novelty may help explain why the body has such a hard time neutralizing dioxin.

Dioxin is a large and very stable compound. It exists, in solid state, as tiny crystals that look like clear or milky white needles. Dioxin's melting point is high, 582 degrees Fahrenheit. Its boiling point, 835.7 degrees Fahrenheit, is just 65 degrees cooler than the surface of Venus. Its vapor pressure, a measure of how quickly dioxin evaporates, is extremely low. In short, dioxin is unlikely to melt, boil or evaporate away. (Though it can be degraded, slowly, by sunlight.) If it gets airborne, dioxin will stick firmly onto whatever particles it encounters in the atmosphere and hitchhike great distances. High concentrations have been found in the arctic where there are no industrial sources of dioxin. Waterborne dioxin adheres just as strongly to sediment particles. Taken together, these chemical properties help explain how dioxin persists in the environment and in the human body. Its half-life in both is measured in years.

Dioxin has other chemical characteristics that make it extremely toxic. For example, dioxin is lipophilic. It loves fat and will accumulate in fatty tissue. This is how dioxin is able to travel up the food chain, from microscopic algae all the way to humans. Levels of dioxin in fish or beef cattle can be hundreds of times higher than the concentrations of dioxin in the lakes or pastures that these animals call home. All residents of industrialized countries carry levels of dioxin in their bodies. More than 90 percent of that load comes from eating beef, dairy products, poultry, pork or fish. The average person will consume between 1.2 and 3.0 picograms of dioxin per kilogram of body weight per day. (A picogram is one-trillionth of a gram). This trifling amount comes perilously close to exceeding the average daily intake range — 1.0 to 4.0 picograms/kilogram of body weight/day — that has been deemed safe by the World Health Organization. Dioxin is dangerous even in trace amounts. The simple act of eating puts us all at risk.

Dioxin has one more troubling property. It is hydrophobic. It doesn't dissolve in water, which means the body can't excrete dioxin in the usual ways, by urinating or perspiring.

It's safe to say that scientists now understand dioxin's chemical nature. What remains elusive are the details of how dioxin behaves once it gets inside the body, how exactly it manages to disrupt so many of the body's normal functions. Scientists know that dioxin generally enters the environment as a liquid or aerosol, and that humans come in contact with dioxin by eating, breathing or touching it. Once dioxin enters the body in any of these ways it can move through the bloodstream and enter the cells of any organ.

Inside the cell, dioxin is intercepted by the aryl hydrocarbon, or Ah receptor. This ancient protein, present in all vertebrates, triggers the breakdown of xenobiotics such as dioxin. It also appears to play a key role in the life and death cycle of cells.

Like the antibodies in our immune system, the Ah receptor is part of the body's surveillance corps. While antibodies intercept biological intruders such as foreign bacteria, the Ah receptor protects us against foreign chemicals. Once an Ah receptor spots a foreign chemical like dioxin, it latches on tightly and escorts its unwanted charge to the cell's nucleus where the process of decontamination will begin.

When it reaches the nucleus, the AhR-dioxin tandem locks onto a specific group of genes and switches them on. The activated genes launch production of specialized enzymes, which in turn begin the stepwise breakdown of dioxin, like breaking down cardboard boxes before tossing them into the recycling bin. But for some reason, our deconstruction/elimination system doesn't work so smoothly with dioxin. Our bodies do eventually manage to break down and eliminate some of it, but the detox occurs at a glacial pace. Dioxin is still present in residents of Seveso, Italy, who were exposed more than 30 years ago. By hanging on, explains Dr. Oliver Hankinson, a professor of pathology at UCLA Medical Center, who has been studying dioxin since 1976, "dioxin stimulates the activity of the Ah receptor in a very strong and sustained fashion."

The upshot is that the Ah receptor remains parked on the cell's DNA and the targeted genes remain locked in the "on" positions. Scientists suspect that over time the Ah receptor may also activate neighboring genes. This dioxin-stimulated Ah receptor disrupts the normal functioning of a cell. In this hyped up state, cells can divide, differentiate and die in disorderly fashion. The over-stimulated Ah receptor also seems to corrupt the cell's mechanism for making proteins. Defective batches of estrogen, testosterone, insulin, adrenaline or growth hormones can start popping up anywhere in the body. It's no wonder Viet Nam veterans exposed to Agent Orange reported such a dizzying array of symptoms and conditions in themselves and their children, from nearly every type of cancer to acute neuropathy, diabetes, porphyria and birth defects.

The plant managers and executives at Diamond Alkali couldn't have known about dioxin's destructive effect at the cellular or molecular level. They couldn't see the faulty enzymes and hormones. They did know what was causing their chloracne problem though, and they had been warned about the potential for far more serious health consequences. Chemical companies in the U.S. and Europe were freely, if quietly, exchanging information and concerns about dioxin. "Tragically," writes toxicologist Ellen K. Silbergeld in the 1993 book *Toxic Circles*, "while industry shared this information within its own circle, industry medical and toxicological staff withheld such knowledge from those most at risk, the workers." No dermatologist or plant manager ever told the crews at 80 Lister Avenue about the true nature of the chemicals they were handling every day.

Despite warning signs, Harry Heist and his fellow workers didn't realize the risk they were taking when, twice each day, they fastened drums of Agent Orange inside the railroad boxcars that carried the containers from the plant site. "We all broke out in big boils," said Heist in his 1983 *Star-Ledger* interview. "[They] looked like giant red pimples and were painful. Some faded away, but others never did. I got them for years after I left the plant, and still do once in awhile. Very tiny blackheads would also appear on our faces. One fellow's face was completely covered with them. We were too young to know any better. We figured we were working with weed killers… We just held our breaths when we had to and put up with the conditions."

Heist also recalled a pungent smell that permeated the 80 Lister Avenue plant and clung to the workers and their clothes. Employees at a neighboring factory reported the smell to U.S. public health officials. Dr. Jacob Bleiberg, a dermatologist hired by Diamond, also alerted U.S. Public Health Service (USPHS) doctors. Blieberg was concerned about the uptick in chloracne cases he was seeing at the Lister Avenue site. He sought out Dr. Donald J. Birmingham, a veteran USPHS dermatologist who had experience with chloracne. Birmingham had investigated a 1949 outbreak at a Monsanto chemical plant in West Virginia.

Birmingham and a colleague, Dr. Marcus Key, inspected the Lister Avenue plant in 1963. Their subsequent report echoed the 1955 findings of Dr. William York. Birmingham and Key recommended renovating, enclosing and cleaning up the Lister Avenue facility. But neither the inspecting doctors, nor the U.S. Public Health Service, nor the state of New Jersey ever followed up. Each assumed that Diamond had complied with the report's recommendations on its own. Speaking to the *Star-Ledger* in 1983, E. Lynn Schall, the director of New Jersey's occupational health program from 1955-70, explained that the state rarely followed up on such reports.

The case that Michael Gordon and the Ironbound community brought against Diamond Alkali—by then the Diamond Shamrock Corporation—was settled, out of court, on Thursday, January 24, 1991. Diamond agreed to pay $1 million in damages to the Ironbound residents, but refused to admit any liability. The settlement came rather abruptly about seven weeks into the trial. Diamond still hadn't called any of its expert witnesses. An earlier ruling by presiding judge Leo Yanoff of New Jersey State Superior Court effectively forced the settlement. Yanoff had dealt the plaintiffs' case a serious blow when he barred any Diamond employee who had pursued a separate workman's compensation claim from participating in the Ironbound suit. An appellate court upheld the decision. Diamond's former workers would settle their separate suit for $1.6 million.

The combined $2.6 million award was far less than the millions in damages being sought. But at that point, said Gordon, "it made sense" for his remaining 72 clients and their families to accept the offer. When I ask Michael Gordon whether he was satisfied with the outcome of the case, he offers a qualified yes. "Given where the Ironbound Community was and the resources it had, and the point in time in the development of these kinds of cases, it was an excellent result. We got a few million and made sure there was a proper recognition of the extent of the problem." In big picture terms, says Gordon, the dioxin case was a wake up call for New Jersey's environmental regulators. In its aftermath, the state's standards for hazardous waste regulation and litigation became far more rigorous and sophisticated.

Arnold Cohen, Gordon's community partner in the dioxin fight, has a different take on the outcome. Arnold is proud of the Ironbound community's mostly successful struggle against waste incinerators. But he sounds, by turns, disgusted and resigned about dioxin. The state, he says, "did the minimum necessary, rather than looking at dioxin as a long-term issue. They didn't go back to retest to see if their cleanup was successful. They didn't do long-term health studies. There doesn't seem to be any interest in whether dioxin is still present. There's no give back to the neighborhood. Some of the workers put up a basketball court with their own money. But in terms of the millions spent for cleanup, nothing went to bettering the lives of the people living in that area. We tried to push for a community center, a health center, something real that compensates folks for all they had to live through and helps alleviate their fear in some way around what they might be facing in the future." Because, notes Arnold, the dioxin story is far from over: "The whole issue of the Passaic River cleanup is still ongoing."

Twenty-nine years after Governor Kean's emergency 1983 press conference, the dioxin that Diamond Alkali dumped into the river is still there. The New Jersey Appellate Court took pains to mention this sad fact in its 1992 opinion from the *Diamond v. Aetna* case. Here again is the Honorable David S. Baime, writing for the court: "We digress to note that neither Federal nor State environmental protection agencies have directed Diamond to remediate the damage to the river. As Diamond correctly points out in its brief, the claims which are the subject of this litigation do not encompass losses resulting from the discharge of substances into the Passaic River. We nevertheless recount this evidence because it bears upon the state of Diamond's knowledge and intent regarding the environmental damage caused by its operations. At least to some extent, this evidence disclosed a less than benign indifference to the consequences of Diamond's operations that directly bears upon whether other discharges and their effects were accidental or inadvertent."

The talk nowadays, and there is lots of it, involves whether and how to safely dredge the dioxin-drenched sediments, and even more controversial, who should pay for it. Michael Gordon is back working on the dioxin issue, helping the state sort through the complex legal issues. I ran into Gordon at the Passaic River Symposium in October 2006.

The event was the second such conference hosted by Montclair State University's Passaic River Institute. The symposium took place at Montclair State's new conference center, a sprawling 7th-floor aerie atop the school's University Hall where floor-to-ceiling windows offered unobstructed views of the Manhattan skyline some 25 miles to the east.

The crowd milling around the registration table that morning was a Who's Who of Passaic River luminaries from the many organizations and agencies that study or oversee or regulate, remediate, rehabilitate, or otherwise have some business with the river. There were several hundred of them — academic researchers, government bureaucrats, federal, state and local politicians, lawyers, journalists, advocates and aficionados. Many were new to their posts.

Colonel Aniello Tortora was making his debut as the new New York District Commander for the Army Corps of Engineers. The Colonel, just six weeks into his tenure, wore combat camouflage fatigues and delivered a let's-all-get-along message.

Alan Steinberg, newly-appointed Region 2 Administrator of the U.S. Environmental Protection Agency, which has jurisdiction over the Passaic River watershed, shared the advice his lawyer son had just given him. "Mother Nature is not a red or blue state constituent," said the Steinberg scion. (Alan Steinberg has since moved on.)

William J. Pascrell, the dapper don of U.S. Congressmen (New Jersey's 8th District), and a veteran of New Jersey and Passaic River politics, touted his environmental credentials, floated a plan to establish a national "water trust fund," and insisted that, by working together, we can restore the Passaic to its former "majesty."

And Lisa Jackson, at the time the fiery new Commissioner of New Jersey's Department of Environmental Protection, and now head of the federal EPA, exhorted the crowd to dredge dioxin hotspots *now*. "By not removing the dioxin, we are as guilty as those who put it there," declared Jackson, accusing polluters of ducking cleanup responsibility by lobbying (successfully) for more study, and daring them to "prove me wrong," by cleaning up their mess.

Rutgers' Robert Chant was scheduled to present his Passaic River sediment findings later in the day. Ella Filippone, executive director of the nonprofit Passaic River Coalition, was also on the program. So was Andy Willner. I spotted Carol Johnston of the Ironbound Community Council in the audience, and of course Carl Alderson, Coastal Restoration Specialist with NOAA and my Passaic River guide, who arrived a little late and without one of the posters he was supposed to present because the large-format color printer at his NOAA office in Sandy Hook was on the fritz and Carl refused to pay the $87 that the guy at Kinko's wanted to charge him for the rush job.

Yes, the Passaic River gang was all there. But surveying the faces in the room made one Passaic River fact painfully clear: there were a lot of voices speaking on behalf of the river, so many people from so many jurisdictions with so many vested interests in so many aspects of the river's health, use and maintenance. But one voice, one person was glaringly absent: the person with the power to enact change. That person doesn't exist.

I step out of my kayak in Newark's Riverbank Park and stand for a moment, stretching at the water's edge. The light is fading fast. Riverbank Park is a rare stripe of Ironbound green sandwiched between the river and busy Raymond Boulevard. The park continues on the other side of Raymond Boulevard too, where baseball diamonds, soccer fields and playgrounds now cover land once occupied by Edward Balbach's refining and smelting company.

Riverbank Park is Nancy Zak's favorite place. "It's one of the few wild spots in the Ironbound where you can get to the river and commune with the water and the trees," she says. Riverbank Park is also the reason Nancy and Arnold are still in the Ironbound. "We were going to do what everyone was doing: move out from Ironbound to Maplewood or Montclair," she says. This was back in 1996 when their daughter Beth was four years old. But before they managed to escape, fate, in the form of one more community cause, intervened.

Nancy and some fellow Ironbound activists got wind of an Essex County plan to turn the 10-acre Riverbank Park into a stadium for the Newark Bears, a minor league baseball franchise that was owned at the time by former Yankee catcher and New Jersey native Rick Cerrone. The idea still horrifies Nancy. "We thought, 'how could they?' Who was behind it?'" she says. "Turned out it was every single politician — local, county, state, even federal — with the exception of one city councilman, who voted with us. Arnold and I ended up staying because we got really pulled into that fight. It was extremely emotional, an all-consuming David-and-Goliath kind of thing."

The battle over Riverbank Park dragged on for seven years. During that time, SPARK (Save the Park at RiverbanK), the group that Nancy, the ICC and other neighborhood activists hastily formed, fought for — and won — the right to have public hearings on the issue. When officials turned a deaf ear to the public outcry, SPARK volunteers collected 4,000 signatures (in a day) to force a citywide referendum on the issue. When the referendum failed to stop the stadium juggernaut, SPARK appealed to the New Jersey Historical Society and, at the eleventh-hour, to the National Park Service. "In the end," says Nancy, "the National Park Service officials were the ones who stopped it."

The Newark Bears did get their stadium, on Broad Street, well upstream of Riverbank Park. The surviving green space, a small gem designed by the sons of Frederick Law Olmstead and finished in 1910, still preserves a glimmer of the once beautiful and mighty Passaic. "The river has always been here," says Nancy. "I think of it as a wild thing that comes through this really over-inhabited land. The Passaic is powerful. There's something about water, I guess. You talk to the old-timers and they used to swim in the river. I just can't wait until we can get out there again."

4 | DIRTY WATER

IT'S WORTH PAUSING HERE FOR A MOMENT to point out that the Passaic is not the only polluted river in America. Nor is it America's most endangered river. That dubious distinction, bestowed each year since 1986 by the nonprofit American Rivers organization, went in 2011 to the Susquehanna, which flows through New York, Pennsylvania and Maryland on its way to the Chesapeake Bay. Eagerness to extract the natural gas locked inside the shale formations beneath the river and its tributaries is threatening to poison the system that supplies drinking water to millions of people.

The Passaic has made the American Rivers' Most Endangered list three times in the last 20 years. In 1990 and 1991, the river was cited, not for pollution, but for the threat presented by two large flood control proposals from the Army Corps of Engineers. The Passaic appeared again in 1998 for the sediment contamination in its Lower Valley. The river has been absent from the list since then.

I suppose it's comforting, in an odd and perverse sort of way, to know that there are waterways throughout the country that are worse off than my hometown river, that are more poisoned, more degraded, more at risk from human intervention and demands. It is certainly true that the lower Passaic is cleaner now than when I was a kid back in the sixties. More and better wastewater treatment plants, and tougher environmental standards for industry have helped to improve the water quality. But it's still dirty water, especially in the river's industrial south end, which makes it all the more distressing that since 1998 at least ten U.S. waterways each year find themselves in more desperate straits than the Passaic.

It's a wonder and a shame to see how many ways we have found to mistreat these critical, beautiful resources. Over the years, the variety of insults threatening waterways on the American Rivers' top ten lists has included mountaintop removal mining (on the North Fork of Montana's Flathead River), coalbed methane drilling (McCrystal Creek in New Mexico), phosphorous from human waste (the Saluda River in South Carolina), cow manure (Tennessee's Roan Creek), toxic sediments (Kinnickinnic River in Wisconsin), radiation (the Columbia River in Washington State), PCBs (New York's Hudson), heavy metals and acidification from mine tailings (the upper Arkansas River in Colorado), oil and gas development (Beaver Creek in Alaska), commercial construction (the Lower St. Croix National Scenic Riverway in Minnesota and Wisconsin)—and then there's Mississippi's Pascagoula River, where the U.S. Department of Energy wanted to store 160 million barrels of oil. Scrolling through American Rivers' Top Ten Most Endangered lists makes it clear that we are a nation of substance abusers, pumping our collective bloodstream full of junk. And all that junk flows downstream.

River pollutants don't recognize state or national boundaries. Just as dioxin drifted from the mouth of the Passaic into New York Harbor, Mississippi River waste, mostly sewage and agricultural runoff from the 31 states along the 2,348-mile-long river, spews into the Gulf of Mexico every day. The onslaught has created a so-called "dead zone[1]" in the Gulf,

---

1 Dead zones also exist in the Chesapeake Bay, off the coast of Oregon, and in freshwater habitats such as Lake Erie.

a New Jersey-sized area off the Louisiana and Texas coasts where low oxygen levels are effectively suffocating marine life in one of the most productive seafood fisheries in the world. That's what my great Grandma Sullivan would have called shitting in the nest.

And toxins are just part of the problem. By building countless dams and bulkheads and highways and office towers across and along our rivers, we steal their water, alter their courses and disrupt their whole intricate ecology. We turn them into robo-rivers.

One of the cruelest and all too common ironies is apparent in American River's 2009 list, where four of that year's most endangered rivers (the St. Croix, Flathead, Snake and Beaver Creek) belong to America's most elite water club: the National Wild and Scenic River System. This rarified designation, created by Congress in October 1968 through the National Wild and Scenic Rivers Act, has been granted to only 11,434 of the 3.5 million river miles in America. That's about one-quarter of one percent of the nation's rivers. We even foul our most magnificent nests. Persistent threats to Wild and Scenic Rivers like the Snake and Flathead show how a broad spectrum of environmental conditions and regard can coexist along the length of a single river, everything from beauty to the beast.

The Passaic River does not enjoy a Wild and Scenic designation. But parts of the upper river are so lovely that they can make you forget about the mess downstream. At quiet times of day, in the early morning and late evening when the light hits the water just so, even stretches of the lower Passaic offer glimpses, ever so fleeting, of the majestic river that once was and perhaps, in time, could be again.

5 GREAT SWAMP TO CHATHAM

I AM STANDING AT THE BOTTOM OF A VAST ANCIENT LAKE. It formed some 20,000 years ago at the end of the last Ice Age, and lingered for thousands more. Thirty miles long and eight miles wide, the lake extended from present-day Far Hills north to Pompton Plains, completely filling the valley that lies between the second Watchung Mountain and the Ramapo Range to the west.

I am standing in the deep end, in Basking Ridge, New Jersey. The surface is 240 feet above me, shimmering in the late afternoon sun. The Passaic River is down here too. Its waters are buried, its character diluted by the tons of ice melt that made this Glacial Lake Passaic. But the river is still flowing, still searching for an outlet to the sea.

I am waiting for Carl Alderson, Coastal Restoration Specialist for the National Oceanic and Atmospheric Administration (NOAA), and my Passaic River paddling guide. We agreed to rendezvous here in the parking lot of the Somerset Environmental Education Center. The Center is at the edge of New Jersey's Great Swamp National Wildlife Refuge. The May air is t-shirt warm and it smells of hay and pond. Oblique rays from the low-hanging sun chisel the landscape, sharpening every object and color:

the sloping green metal roof of the Environmental Center; the lone wood duck nesting box nailed to a pole in the middle of the pond; the two portly birders trundling across the cedar boardwalk towards the swamp. It's an Edward Hopper painting without the loneliness.

Environmental Center workers are leaving for the day. They wave and call out goodnights to each other as they fan out across the parking lot to their cars. In a stab at reconnaissance, I fall in step beside one middle-aged supervisor and inquire about water levels in the Passaic. He warns against paddling upstream—too shallow this time of year. But despite scattered reports of logjams downstream, he thinks a paddle in that direction should be doable.

I thank him and wave as he gets into his car. The lot is almost empty now. Dust from the exodus hangs in the air.

Before long, a dented maroon Camry with two orange kayaks on top comes zipping into the parking lot, and out pops Carl. He smiles and waves. "You ready?" he yells, then takes to rummaging through the trunk of his car.

Today, Friday, May 5, 2006, is the beginning of our kayak trip down the Passaic. We're allotting ourselves four full days on two separate weekends to paddle the river's navigable length, roughly 75 miles.

Carl is taller than I imagined, 6'3", and rangy, a former high school shortstop. His light brown hair is cut short and just beginning its inevitable journey aft. He reminds me of a taller, leaner, less smirky George W. Bush.

Carl works for NOAA's Damage Assessment Remedial Restoration program. He uses settlement money from oil spills and other industrial releases to restore coastal habitat. His office is in Sandy Hook, New Jersey. His territory extends south to the state of Delaware and north to the Canadian border. That's a lot of real estate, a lot of restoration.

Carl is a closer. NOAA hired him to clear the agency's oldest and most vexing projects. "My job," he says, "is to basically make it happen." Since joining the agency in 2003, Carl has taken many projects off NOAA's books, including salt marsh restorations in Woodbridge (17 acres) and Jersey City, New Jersey (42 acres). Once the official Passaic River damage assessment is finally complete, Carl hopes to manage the Passaic River projects that result.

Despite the late hour, Carl is in no hurry. This relaxed approach will prove exasperating at times for an impatient person like me. But I will come to admire and depend on Carl's talent for living in and savoring the moment.

We have a few chores to attend to before we can hit the road this evening. First, Carl attaches his extra set of roof racks to my rental car. Then it's time for "orientation." He reaches into the Camry and produces two clear plastic folders. He hands one folder to me and spreads the contents of the other across the dusty hood of his car. There's a detailed trip itinerary inside, plus a gear checklist, laminated river maps and a graph showing the most recent water depths recorded by NOAA gauging stations downstream at Millington and Chatham. The river, notes Carl, is running a little shallower than he'd like. We won't, as a result, be able to kayak upstream from the Environmental Center.

With the map as visual aid, Carl explains that tomorrow we will paddle the last few navigable miles of the Upper Passaic and on into the river's mid-section. The day will begin right here in Basking Ridge. We'll put the kayaks into the river at a boat launch used by local fishermen. It's back down the gravel road, about a quarter mile from the parking lot. From there, we'll paddle almost due south, hugging the southwestern edge of the Great Swamp for about two river miles. We'll exit the Refuge at, or rather through, the Millington Gorge. After we shoot the Gorge, we'll head north with the river, paddling into the Passaic's Central Basin, past the quiet little communities of Stirling and Gillette, and then the slightly larger and more sophisticated towns of Berkeley Heights and New Providence. We'll end the day in Chatham, about five overland miles northeast of our starting point.

The trip looks pretty short on my laminated map, maybe three inches. It's actually 19 river miles. Carl and I discuss the implications of low water along the way, especially through the Millington Gorge and again on the riffled approach into Chatham. The upside: no need to wear the dorky yellow helmets I saw wedged inside the Camry's trunk. The downside: we might have to portage the boats.

It's dusk by the time we set out for Chatham, the endpoint of tomorrow's paddle. We'll leave my boat-racked rental car there overnight. We make two stops on the way. The first is on the one-lane steel bridge that crosses the Passaic at White Bridge Road, barely a quarter-mile from the Environmental Center. Carl wants to have a look at the river. Or, perhaps he wants *me* to take note of this serene beginning part of the Passaic.

The bridge sits low, maybe four feet above the surface of the water. With night sounds all around us, we walk out on the iron grate to the middle of the bridge and lean against the waist-high rail. Its aging white paint, what's left of it, is chipped and flaking and almost completely consumed by rust. The river, a mere creek at this point, slips silently beneath us. Its surface is dark and very still. The first time Carl kayaked the Passaic, in 2003, some of his fellow paddlers went swimming here. "The river tends to be slow and shallow in this area," he says. "But it'll pick up speed once we get farther down."

Back in the cars, I follow Carl north across the wide, flat basin of the Great Swamp. I spot a red fox in the flatlands, grooming itself in the front yard of one of the small split-level homes we pass. A few miles farther on we start up Long Hill, one of the broken ridges that makes up the westernmost range of the Watchung Mountains.

The Watchungs are three parallel ridges pitched in a northeasterly direction and curved like sly smiles. They stand between Mendham and Newark. The westernmost Watchung, the one Carl and I are scaling, is a series of interrupted crests. Its two sisters to the east are solid rock walls that extend some 40 miles, from Somerset, New Jersey, north to Paterson.

The name Watchung, or "Wacht unks," means "high hills" in the dialect of the Munsee Indians, a branch of the Lenni Lenape tribe. The Watchungs aren't especially dizzying. High Mountain in Wayne is the pinnacle at 879 feet; the other ridges top out at 500 feet. But the Watchungs are formidable. They began as molten lava that bubbled up from the earth's core 175 million years ago. When the lava reached the surface it cooled and hardened into igneous basalt, a volcanic rock that is dark gray in color, rich in iron and magnesium, smooth in contour and very, very hard.

When the southward-flowing Passaic meets this basalt barrier—just beyond Millington—it has no choice but to perform an abrupt about-face. The river turns sharply north and enters the Central Basin, feeling its way along the western face of the middle Watchung ridge in search of an opening. It finds one some 40 miles downstream (or north) at Little Falls, and then another wider pass through the easternmost Watchung ridge five miles later in Paterson. Having burst free of the mountain fortress, the Passaic turns south again, flowing down the length of the Lower Valley and on into Newark Bay.

Carl pulls over at the top of Long Hill ridge. We get out of our cars and walk to the edge of a wide pullout on Fairmount Avenue and turn to the southeast. A line of high-tension towers marches up and over the ridge from that direction. "That's where we'll be tomorrow," says Carl. He points down into the steep-sided valley that lies between the Watchung ridge we are standing on and its sister to the east. I can't see to the bottom of the valley; the vegetation is too thick. But I know the Passaic is down there, some 500 feet below us, snuggled between the two Watchungs, making its way north.

Over dinner that night at Chatham's Taste of Asia restaurant Carl explains what a Coastal Restoration Specialist does, and how his role fits into the process that NOAA and its fellow regulatory agencies follow in the wake of any environmental accident or insult. "First, you respond and contain," he says, between bites of Chow Yang chicken. "That's the emergency response. Second, you assess the damage and determine how it can be compensated. Finally, there's me, compensating, performing the work."

The emergency responders, personnel from various state and federal agencies who oversee the actual cleanup, are led by the U.S. Coast Guard and Environmental Protection Agency. Assessment teams from NOAA's Damage Assessment Center work with the U.S. Fish and Wildlife Service to compute the extent of the environmental insult. It's a complex calculation.

Environmental damage is measured in Discounted Service Acre Years (DSAY). "We call them dee-says," says Carl. "And it's not just acres. It's the services provided by coastal wetlands and waterways that were lost over the time that the area was impacted. I'm given the calculated number of Discounted Service Acre Years, and I help determine that a project of a certain size will afford a certain number of DSAYs. So for instance, I might have a suite of projects available that includes a salt marsh, a floodplain, a riparian border and a shellfish bed. Which ones do we need? And how many acres of them are necessary to make the public whole?"

The current DSAY model wasn't in use in 1990 when Exxon's Linden, New Jersey oil refinery sprang a leak. But it's a safe bet that the so-called Exxon Bayway spill, an environmental disaster that launched Carl's career as a restoration specialist—and Andy Willner's as an advocate—would have generated lots of Discounted Service Acre Years.

Exxon's massive Bayway facility covers 1,300 acres of industrial land between Linden and Elizabeth, New Jersey. Since 1909, the refinery has been turning crude oil into gasoline, home heating oil, jet and diesel fuel, and other petroleum-based products. The spill occurred when a four-foot-long gash opened up in the 6.7-mile-long pipe that connects the refinery to its distribution terminal in Bayonne. The pipeline runs underneath two waterways—the Arthur Kill and Kill Van Kull—and across acres of wetland. The spill occurred on January 2, 1990. Carl was working for the New York City Parks Department at the time. "I always said it was a little too much eggnog out at the Exxon plant," he says, only half kidding. "The pipe had been rupturing and the alarm had been going off for weeks, but they thought it was a false alarm, so they switched it off in the middle of the night, and nearly 567,000 gallons of #2 heating fuel leaked out into the waterways and all over the marshlands." Estimates about the extent of the damage ranged from just a few acres of marsh to 250, depending on who was doing the counting. But in the final tally, says Carl, "hundreds of acres were affected, 11 acres were wholly destroyed, and the fragile ecology of the Arthur Kill was devastated."

The oil never reached the Passaic. The river's mouth lies several miles to the north of the ruptured pipeline. But damage to the estuary was extensive. "At the epicenter of the spill—at Old Creek Place in Staten Island—roughly five acres was irretrievably lost," says Carl. "There's a very toxic component to refined oils. It's a phytotoxin, a plant killer. So we started primary restoration, rebuilding marshes in the patches that had been denuded by the spill. When we ran out of primary restoration areas we did compensatory restoration. That's where you have to go off site to compensate for the loss." The newly-completed Lincoln Park restoration in Jersey City was just that: 42 acres of displaced compensation to help make amends for the eight miles of New York Harbor shoreline that got slimed in the Bayway disaster.

Twenty years and 47 restored acres later, Carl is still compensating for the Bayway loss. Most of his current caseload comes from that one 1990 leak. He oversees all the Exxon Bayway restoration projects for NOAA. Carl hopes to have active Passaic River projects as well. He's waiting on damage assessors to translate two centuries of environmental destruction into Passaic River DSAYs. Until then, he says, "all I can do is identify the opportunities and track them and just gear up, get ready." Paddling the Passaic is part of getting ready. It's a good way to spot potential restoration sites. Which makes our kayaking trip, for Carl at least, a kind of busman's holiday. Carl, the closer, is also a location scout.

We make it back to the Great Swamp put-in by 9:30 the next morning. There's a spring chill in the air and a menacing wall of dark clouds to the southeast. Neither will last. We unload the kayaks— two, 14-foot-long hard plastic Pungo140s— and carry them down the short dirt path from the put-in's gravel parking lot to the river's edge. The low bank is firm and dry with thick patches of short green grass. We set the boats down on the grass a few feet from the actual launch point, a rounded dirt wedge that slopes gently down to the water. We stow the gear. We review the maps. We snap photos of each other standing next to our Pungos, wearing our yellow life vests and fingerless paddling gloves. At last, we launch the boats.

I let my big orange Pungo drift downstream for a while, getting a feel for the speed of the current, which is lazy. The river is narrow and gentle in the Great Swamp part of its run. This is a Passaic I do not know. The channel is about 30 feet wide, bounded by emerald green marsh on one side and by tawny bottomland woods on the other. The water is shallow and dark and clear— inviting. Carl and I glide beneath the low bridge at White Bridge Road, the one we stood on last night. I reach up and touch its crusty underside. The bridge must be easily swamped at high water times. We continue south, following the shallow loops that the river makes as it forms the southeastern boundary of the Great Swamp. We remain within the Great Swamp refuge for the first two miles. It's a calm, slow, Zen paddle. We don't talk much. Several small tributaries trickle into the Passaic along the way: the Black, Great, Loantaka and Primrose brooks. We hardly notice them.

At river mile 1.7, as I prepare to paddle under the bridge at South Maple Avenue in Basking Ridge, I hear Carl call out from behind me. I bring my boat around. He is gesturing downstream with one hand and waving me back towards him with the other. Millington Gorge must be just ahead. I paddle back to Carl, who has pulled his kayak up to a gauge station on the riverbank to check the water depth.

The Millington Gorge is a narrow cobbled chute through the broken hills of the westernmost Watchung ridge. This is where the Passaic exits Great Swamp. The gorge was carved, or rather worn away, some 20,000 years ago by the overflow waters from Glacial Lake Passaic.

Millington Gorge is not a spectacular sheer-sided vault like the gorges out west. Its steep banks are forested for one thing—eastern redbud, shagbark hickory, river birch, and splashes of pink from the occasional dogwood tree and viburnum bush. But the Passaic runs downhill through here and it runs fast. When Carl first kayaked the Gorge in June 2003, the river was at flood stage. Heavy rains, unusual for that time of year, had whipped the drowsy Passaic into a torrent. "That was a thrill ride," says Carl. "You were just getting your river legs when you were thrown into this shotgun."

The early morning clouds have given way now. Shafts of sunlight poke through the canopy and dapple the riffles before us. I watch a red-bellied sapsucker scale the trunk of an eastern redbud tree. A towhee and a common yellowthroat are singing somewhere close by, well concealed in the understory. The gauge station is a square wood-shingled shack at the water's edge. It looks like an outhouse. "4.8," says Carl, reading the gauge. That's about an inch below the minimum depth that New Jersey paddling guru Edward Gertler recommends for passage through the Gorge without "dragging." Gertler's book, *Garden State Canoeing: A Paddler's Guide to New Jersey*, is the paddler's bible.

We proceed anyway and learn pretty quickly that Gertler is right about the water depth. The current whisks us into the Gorge. Our boats bump and scrape along the stones on the river bottom. At times, the Pungos simply grind to a halt, beached on a cobble or two, and no amount of paddle poling or hip thrusting will dislodge them. At one point, the very center of my boat gets caught up on a sharp rock and the Pungo begins to rotate in the current like a compass needle. A father and his two young daughters playing on the steep bank wave in sympathy as Carl and I step out of the kayaks and maneuver them into a deeper part of the channel. We are forced to perform this operation at least a half dozen times through the mile-long Gorge. Visions of a "thrill ride" fade after the third time I climb out of my boat. Carl and I didn't run the Millington Gorge this day; we more or less walked it.

Still, it's something of a miracle that we are even on this upper part of the Passaic River. Most of New Jersey's wild spaces have come under siege at one time or another, and the land in this part of the Passaic River Basin is no exception. In 1959, the New York Port Authority tried to turn Great Swamp— all 7,600 acres, along with chunks of bordering Chatham, Long Hill, Madison and Harding— into an international airport. It took ten years and a lot of lobbying before local citizens finally foiled the airport plan. They convinced Congress to declare the Swamp a National Wilderness Area, which put the land permanently off limits to development of any kind. In 1968, the U.S. Army Corps of Engineers proposed damming this whole stretch of river and flooding the Millington Gorge. The idea, to essentially recreate a portion of Glacial Lake Passaic, was one of several flood control solutions floated by the Corps after Congress directed it to study the Passaic River flooding problem in 1936. Local activists ultimately defeated the dam proposal and saved the Millington Gorge for paddlers like Carl and me.

The battle for the Gorge gave birth to the Passaic River Coalition (PRC), the oldest and most well known Passaic River advocacy group. It also launched a decades-long public debate over what to do about the Passaic River's penchant to overtop its banks.

Despite today's low water, hydrology experts agree that the Passaic is one of the most flood prone river systems in the U.S. The Federal Emergency Management Agency (FEMA) ranks New Jersey among the top four states in the country when it comes to flood-related property loss; the Passaic and its tributaries are to blame. For more than 150 years, public officials and their floodplain constituents have tried to contain the river, digging ditches through wetlands to improve drainage, widening and dredging the river's channel to ease flow, and constructing dams and reservoirs to keep storm water at bay. But the Passaic just keeps on flooding.

The geology of the river's basin is a principal cause of the flooding. The development in the river's floodplain is the reason the flooding is so costly.

Every time the Passaic transits the Watchungs— first at Millington Gorge, then farther north at Little Falls, and finally at Great Falls in Paterson— the flood risk rises. None of the mountain gaps is wide enough to accommodate large volumes of water, the kinds of flows generated by storms and by the annual spring melt. When high water chokes these funnels, the river backs up and spreads out into its broad floodplain and surrounding wetlands. Which is as it should be. Floodplains and wetlands exist to absorb excess water. Except that three million people live in the Passaic River Basin. The floodplain and most of its adjoining wetlands are chock full of housing and industrial developments. The Army Corps of Engineers puts the number of homes and businesses in the Passaic River floodplain at more than 20,000. So when the Passaic rises up to reclaim its floodplain, river water inundates backyards and basements and factories all across north central Jersey.

Since 1900, Passaic River floodwaters have killed 26 people and caused $4.5 billion dollars in damage to homes, businesses and the environment. (Those figures don't include deaths and damage estimates from the 2011 flooding caused by Hurricane Irene and Tropical Storm Lee, which was the fifth worst on record.) Every flood for the last 40 years—in 1968, 1971, 1972, 1973, two in 1975, 1984, 1992, 1999, 2005, 2007 and 2011—was severe enough to trigger a Federal Disaster declaration. During the April 2007 episode, the third worst flood on record, rescue teams evacuated 5,000 people in 11 counties. Things got so bad that year that the Internal Revenue Service extended its April 15th filing deadline for Passaic River floodplain residents.

Carl and I leave the Millington Gorge behind and follow the river around a sharp elbow-shaped bend. This elbow is the southernmost point in the Passaic's zigzagging course to Newark Bay. Once the river rounds this bend below Millington, it heads northeast into the Long Hill Valley and Carl and I find ourselves paddling through a dark, secluded patch of forest just south of Gillette.

This "V" of Passaic, upstream yet south of Chatham, is vastly different from the river I grew up along. It's much narrower, for one thing, and less populated and cleaner and far more natural. This is especially true of the river's western flank, the Morris County side, home to the spacious Great Swamp. This stretch of Passaic forms the east-west border between Union and Morris counties, and a portion of the north-south border between Morris and Somerset County to the south. As Carl and I progress downstream, the river will become more degraded, and the landscape surrounding it will grow more crowded and distressed. But here, at the beginning, the Passaic and its banks still seem young and fresh and more or less untamed.

Much of Morris County remains semi-rural. In contrast to the regiment of cities and towns that line the lower Passaic, there is space between these upriver settlements. Many of the small river towns in Morris County aren't really towns at all. Millington (pop. 3,144), Stirling (pop. 2,307) and Gillette (pop. 3,251)[1] are actually unincorporated villages— Meyersville is a hamlet— within larger Long Hill Township,[2] a narrow 12 square miles of former farmland sandwiched between the Passaic and the Great Swamp. Even Basking Ridge, white-collar headquarters of Verizon and Barnes & Noble, and nearly twice the size of Long Hill, remains an unincorporated part of Somerset County's Bernards Township. Morris and Somerset counties are among the top ten wealthiest counties in the U.S. The living is a little easier on the upper Passaic.

Several hundred yards into the woods near Gillette, Carl and I encounter our first obstacle: a downed tree across the channel. We're probably three miles downstream from the Great Swamp now. This must be one of the logjams the supervisor at the Environmental Center warned me about. The river is still narrow. Fallen trees don't have to be that tall to span the channel. This one is a strapping maple with a substantial trunk and a matrix of branches that spreads up and out across the water like a spindly claw.

Carl calls these fallen trees "strainers," because their branches effectively strain leaves, twigs and just about every other kind of flotsam out of the current. It's not unusual for riverside trees to topple. Current can steal the bank right out from under riparian growth. Bends in the river

---

1 Source: U.S. Census Bureau, 2010 Census.

2 In New Jersey, townships are the oldest form of municipal government. They are typically run by a small elected committee, which appoints one of its members to serve as mayor or township executive, a largely ceremonial post.

accelerate the process, and the Passaic is pretty twisty. One 19th Century geographer called it the "most crooked" river in the state.

Crooked rivers are tough on trees. When a river rounds a bend, the water moves faster on the outside, or concave side of the meander. It's like my Girl Scout troop marching in North Arlington's annual Memorial Day parade. Every time we had to negotiate a curve or a corner in the parade route, the Scouts on the outside had to march a lot faster than the Scouts on the inside in order to keep our parade lines straight. The same kind of speed differential is at work in a river, and it sculpts the river's channel. The faster outside current scoops soil from the outer bank. That soil settles out along inner banks where the water, like the inside Scouts, moves at a slower pace. Over time, this erosion-deposition action actually shifts the riverbed sideways in the direction of the outside bank. The scouring can make casualties of a tree, or any vegetation rooted there.

We've passed several trees whose root balls were almost completely exposed. Eventually, most of them will lose their footing altogether and wind up in the river. The downed and soon-to-be-toppled timber threatens to make this part of the Passaic all but impassable, which is a pity because it's a lovely, quiet stretch of river. I'm beginning to understand why canoe companies don't operate on the Passaic. There's not a lot of access to the river, and no one seems to be keeping the channel clear.

We'll have to portage the boats around this maple log. We pull over to the flattest stretch of riverbank we can find. The ground is carpeted with large maple leaves that are dusted with dried mud, evidence that the river inundates this woodland. Carl jabs his paddle down into the leaf bed to test the firmness of the bank. There's an ankle-deep layer of mud beneath the leaf cover. The mud gets thicker and wetter with proximity to the water line. Getting out of the boats is going to be tricky. But Carl has a plan.

We back paddle across the river, point our Pungos directly at the muddy shore, then paddle towards it as hard and as fast as we can. The momentum carries the bow of each Pungo well up onto the bank. We crawl forward out of the cockpits, across the bows to the very end of our beached kayaks. We step gently onto the bank. We still sink to our ankles. But we'd have been up to our knees if we had parallel parked the kayaks against the bank and tried a more conventional dismount.

To save some time, we decide to carry both boats at once. Carl takes the bows; I grab the sterns. Each Pungo weighs about 70 pounds. They're heavy, but the load is balanced. We lug the boats about 30 yards, through the soggy leaves, over branches and around trees and stumps, to a spot on the downstream side of the strainer that is sort of flat and kind of dry. Carl shoves me off cleanly, but he slips getting back into his boat and his left leg sinks thigh-deep into the muck. With some effort, and a soft sucking sound, he manages to extricate the leg and fling his muddy self back into the Pungo. All in all, we waste about 30 minutes getting around the strainer.

After our third muddy portage, Carl and I make a vow: we are not getting out of our boats again. We'll find some other way around whatever obstacle we encounter. That pledge is put to the test almost immediately as we round another bend in the river and see another massive tree collapsed across the channel. We survey the stout, branching log that lies between us and Chatham, still a good 12 miles downstream.

"Okay," says Carl, finally. "Watch this."

He brings his boat around parallel to the only branchless stretch of trunk. The open space is about four feet long, 18 inches wide. A single spiral of poison ivy twirls around the log like a boa. Using his paddle as a brace, Carl hoists himself out of the kayak and onto the trunk. He reaches down, grabs the Pungo, yanks it across the log and slides it into the water on the other side. Then he lowers himself back in.

"Come on," he says to me.

I make it out of my boat and on to the log, no problem. It takes two tugs, but I get the Pungo up and over and down into the water on the other side. Ignoring the poison ivy that is lashing my right forearm, I reach down and maneuver the boat parallel to the log. But as I pull the Pungo closer for my remount, I lose my balance and tumble backwards into the Passaic.

I was standing on the Rutgers Street Bridge when the fear of falling into the Passaic River first gripped me. Rutgers Street was our neighborhood bridge. It spans the river at the south end of North Arlington. The bridge was clearly visible from the Homelite field. We drove across it all the time. But on this particular summer afternoon, my older brother Johnny talked me into walking across, something I had never done before. I was seven.

The original Rutgers Street Bridge, erected in 1790, was a wooden toll bridge. It was swept away by a flood in 1841, and rebuilt two years later. But the Rutgers Street Bridge we all knew came along in 1914, the same year the Newark Meadows began its disappearing act. The 1914 edition was a two-lane, black steel structure, designed and built by the Strauss Bascule Bridge Company of Chicago. It was one of 28 bridges that crossed the lower Passaic. All of them, with the exception of the 200-foot-high Pulaski Skyway span in Newark, were designed to open for boat traffic.

A bridge can open in three ways. It can swing open like a gate. It can lift straight up like an elevator. Or it can tilt up, from one side or the other, or from the center. The Rutgers Street Bridge tilted. It was a "bascule style" bridge, bascule being the French word for "seesaw." Bascule bridges tilt up from one side, like a sawed-off seesaw; or, in the case of the Rutgers Street Bridge, a sawed-off seesaw with a giant weight on the short end.

The Rutgers Street counterweight was a hulking cement block that was suspended from the girders on the west bank.

The bridge was still opening for boat traffic when we were kids. The Passaic had ceased to be the main avenue for transport and commerce; most goods entering New Jersey arrived at the Port Newark/Elizabeth Marine Terminal and worked their way north and west by truck or rail. But the river channel was still navigable, and the occasional cargo or pleasure boat could still be seen motoring up or downstream, necessitating a bridge opening and the traffic backups these openings occasioned.

Sometimes, our station wagon would get stopped close enough for us to step out and watch the process up close. The rough iron grate which was the bridge's roadway would unhinge itself from the eastern bank and begin a measured sweep upward until it came mount to the sky and loomed before us, a porous monolith, hoisted and held at an impossible angle.

When I finally followed my brother out onto the bridge grate that afternoon, I tried not to look down. But halfway across I stopped and leaned over and peered through the metal honeycomb of roadway. As the river passed between and around the bridge pilings, its dark surface began to swirl and eddy, welling up in some places, curling down in others, in what my Italian grandmother used to call a "rolling boil." The patterns were chaotic, hypnotic. I tried to imagine what was creating them, what undercurrents were roiling the river's surface, and what it would be like to plunge down into all that madness.

The Rutgers Street Bridge was the umbilical cord that connected us to my mother's side of the family. We drove across it on a regular basis to visit her sprawling clan, a jolly band of Irish, English, French and Germans scattered around the western bank of the Passaic in Belleville and Nutley and Bloomfield.

My grandfather, John Westlake, was happiest when his very large brood—four children, their spouses and 23 grandchildren—filled his very modest two-bedroom railroad flat at 39 DeWitt Avenue in Belleville. Over my grandmother May's weary and mostly halfhearted protestations my grandparents would host huge family dinners throughout the winter, spring and fall and even bigger backyard picnics in the summer. My grandfather and his sister-in-law, my Great Aunt Teresa, embraced the summer festivities with gusto and delight and a special brand of Anglo-Irish bonhomie that, after a few cocktails, achieved a kind of effervescence.

In the lively din of one DeWitt Avenue picnic I overheard Aunt Teresa say something about a local parish priest who, as a teenager, had been caught skinny-dipping in the Passaic. This was a startling disclosure, horrific and extraordinary. The fact that a person had actually swum in the Passaic River represented a heretofore unimagined and unimaginable possibility, and my first inkling that the river may have had another life; that there were people—like that priest, like Aunt Teresa—whose relationship with the Passaic was fonder and less troubled than my own.

After that, I sometimes imagined swimming in the Passaic with my brothers and sister and 18 cousins. Water was the one thing missing from the DeWitt Avenue picnics. There was no kiddie pool in my grandparents' backyard, no swimming hole nearby, no place to cool off on those sweltering summer days. I would picture us kids piling into one of the station wagons—every family owned one—and some designated uncle driving us the quarter mile down Rutgers Street to the river, and the lot of us gushing out of the car and into the Passaic. But that was just a daydream. In reality, we'd cool off by dunking our heads into the brain-numbing ice water of the beverage tub.

Carl doesn't snicker or shudder when I slip off the poison ivy log. I land on my feet, waist deep in the river. No need for Carl to break out the rescue rope he packed. Falling in isn't so bad really. We're well upstream of any sewage outfalls. The Passaic is still fairly pristine. The water is pleasantly cool, the river bottom surprisingly firm. I feel baptized, or maybe exorcised—I survived a plunge into the Passaic!—and definitely cleaner. The dip dissolves the river mud that was caked on my legs and sandals. I hop back onto the log and into my boat and off I go, no worse for the dunking.

Fallen trees seem to lurk around every bend. During the next mile or so of river, Carl and I encounter more than half a dozen of these strainers. It could be that suburbanization in this part of the watershed is accelerating the natural channel sculpting process. Paved streets and parking lots can't soak up rainwater like woodlands and wetlands. We never got water in our Bathurst Avenue cellar until new neighbors moved in next door and paved their backyard. Once the concrete replaced the grass we were bailing out the basement after every downpour.

As development encroaches on the upper Passaic, it disrupts natural runoff patterns. Storm water that once seeped gradually into the river from various points along its length now courses off hard surfaces, collects in culverts and flushes into the river at select locations and at high speeds and volumes. These pinpoint spikes in flow are like flash floods that gouge the channel. With the loss of more empty space, this destructive dynamic gets more frequent and more pronounced. The riverbank erodes and trees topple and that presents challenges to area homeowners and to paddlers like us.

Fortunately, Carl is resourceful. The poison ivy portage is just one of three strainer-busting techniques that he improvises in order to avoid hauling our boats around these Passaic River logjams. I dub that first move the Up 'n Over. It did require getting out of the boats, so I suppose, technically, it was a violation of our vow. But we were out of the boats ever so briefly, and as it turns out we will only use the Up 'n Over that one time. We never ran into another tree trunk that was as wide or as branchless as that first maple.

My method of choice is the Teeter-Totter. We use it to negotiate logs with a relatively small circumference and a gap between branches that is wide enough to poke a Pungo through. Like most of our strainer approaches, the Teeter-Totter begins by paddling full tilt toward the opening in order to drive the Pungo's whole front end up onto the log. Using the log as the fulcrum, we then creep forward from the cockpit until our weight tips the bow side down into the water. Then we return to the cockpit and hip thrust the stern in too.

The Jam-o-Rama is the technique we use most often. It is Carl's favorite. I think it appeals to his demolition derby side. It works like this: I go first, paddling headlong into what we determine to be the least dense part of the obstruction. It's like kayaking through a hedge. My goal is to jam my Pungo as deep into the strainer as I can get it. Then Carl rear-ends me, ramming my boat through to the other side. Sometimes it takes two or even three tries. Once my boat is clear, Carl pulls himself through the thicket by grabbing hold of branches or logs or whatever he can reach with his spidery long arms. We aren't scoring any style points with these strainer crossings. But we are staying in our boats and moving steadily downriver, and those are our goals.

So far, none of the strainers we've passed has snagged any manmade debris. That will change as we continue downstream. Soon, every logjam will become a collection point for the discarded refuse of suburbia. Sometimes, we linger at these suburban strainers. Like archeologists, we poke through the garbage trapped in the tangles. The last strainer has it all: plastic bottles, grocery bags, beer cans, car tires, Big Mac wrappers, topless Styrofoam coolers, shards of Styrofoam packing material and every kind of ball you can imagine. Soccer balls, basketballs, tennis balls, whiffle balls, footballs, little squishy sponge balls in bright primary colors, and big dimpled balls of dubious purpose and appeal all bobbing, in orderly flocks, within the strainer's woody embrace. They look like racked billiard balls. Carl collects one of each.

This stretch of Passaic curls through and around some of central Jersey's tonier towns. But the refuse is one of the few signs of habitation we see along the way thanks to the Morris County Park system. The nearly contiguous, 711-acre pipeline of jigsaw-shaped county parks begins at the Route 531 bridge in Millington and extends north to Central Avenue in Long Hill Township. It forms a cylinder of green which wraps the river like insulation around an electrical wire, and creates an illusion of wilderness. Very few houses are visible from the river, and hardly any residents. Nearly three million people live in the Passaic River Basin. We've been paddling for three hours now and we've seen exactly five: the dad and his two daughters in the Millington Gorge, and a pair of trout fishermen a little farther downstream. The New Jersey Department of Environmental Protection stocks the upper Passaic with trout each spring. Needless to say, the anglers weren't too pleased with our intrusion.

It feels strange to say this, but I'm disappointed by the scarcity of people along the river. This upper part is really quite beautiful, strainers and all. I envy the natives their quiet, non-toxic Passaic. Yet hardly anyone seems to be out here enjoying it. Andy Willner once described how his Baykeeper organization tries to convert ordinary citizens into Passaic River advocates. "It sounds like a cliché," said Andy, "but the way we've been most successful is to get them cold and wet and dirty. Get them out in kayaks. Get them picking up garbage, planting *Spartina*, monitoring water quality. Whatever it takes to get people reconnected to the water's edge." Andy's group targets the lower Passaic, and with good reason. But residents along this upper stretch could use a little reconnecting too.

In the winter of 2006, I went along on a scientific cruise up and back down Brazil's Rio Negro River, one of the Amazon's main tributaries. The Rio Negro is named for the color of its water, which is darkened by the tannins that leach in from the surrounding vegetation. Tannins are bitter compounds found in the bark, stems and leaves of certain plants. They stain the main channel of the Rio Negro an inky black and some of the river's small feeder streams the color of blood.

There were 22 passengers aboard our ship, the *Victoria Amazonica*, a 60-foot wooden vessel that looked like an old Mississippi River boat minus the paddle wheel. The manifest included a tropical vine specialist, an E.R. doc, two software CEOs and a couple from Fruita, Colorado who made plaster casts of dinosaur bones for museum exhibits. All were family members or friends or, like me, new acquaintances of the Denver paleontologist who organized the trip. We traveled the Rio Negro for nine days. We saw giant iguanas and pink dolphins and spiders with hot pink feet and tropical birds that looked like flying rainbows. And all along the way we saw people. River people. Out tending gardens beside simple wooden shacks, or fishing or building boats or washing clothes at the water's edge, or passing us in slim canoes and in skiffs with small outboards and in bigger working boats with canvas canopies that shaded fruit or gear or garbage. Sometimes, they tied up to the *Victoria Amazonica* to sell us fish, chickens or fruit, or to gossip, or to negotiate passage up or downriver. The Rio Negro was their highway and their home. It felt well traveled and well loved. Lived in. Lived *on*. The way the Passaic River used to be, when towns along the lower Passaic had rowing clubs and weekend regattas drew crowds of cheering spectators to the riverside.

At river mile six, just past Gillette, the Passaic widens. Carl and I paddle out of the shadows and into a sunny river lake. A few small houses, the first we've seen, dot the southern shore. The opposite bank is forested and empty. Still no people about. The late morning sun is warm and bright. We have the river lake to ourselves. We make our way to the far end where a wide, wedge-shaped sandbar has formed in the middle of the channel. The sandbar is covered with spatterdock (*Nuphar variegatum*), an opportunistic water lily whose thick, heart-shaped leaves have transformed the sandbar into a verdant island.

"It's a pretty odd formation," says Carl. He reconnoiters the perimeter of the island, whose existence he considers evidence "that the Passaic is a low flow river most of the time. When the river widens like this it loses velocity, and it drops sediment out at the end of that low flow area." The sediment that accumulated here formed this mid-channel shoal, which became a spatterdock island dam, which widened the river behind it into the pleasant little lake we're floating along.

Strainer hell, I hope, is behind us. The river up ahead looks wide and sunny too. Trees, all of them upright, line the banks. I can see a baseball diamond through the woods and hear the shouts and cries of Little League. An outlaw band of players in red uniforms dashes down to the river to watch us drift by. We call out to them. They respond with nods and furtive waves, quick wrist flicks from the waist. I stop paddling and watch them watch us until they grow bored and wander back to the field.

There's a dreamlike quality to this paddle down a river that I had so expected not to like. The trees are still leafing out. Their emerging canopies, vibrant in new spring green, are delicate and feathery. Seedpods fill the air, floating through the balmy spring afternoon like dust motes through a shaft of sunlight. There are yellow-green caterpillar-shaped pods from the river birches, rotoring helicopter blades from the silver maples, and from the American beech trees, tiny round balloons of the palest green, with strings attached. The pods are aerodynamic and hydrodynamic, the perfect vehicles for dispersing the seeds of these riparian trees. Riding invisible breezes, they drift and twirl and dip and spiral down onto my hat and my boat and onto the river where they hitchhike on the current, carouseling in the eddies, forming thick chartreuse carpets behind the snags, drifting downstream like so much fecund dust, hoping to find purchase.

Carl and I take our lunch on the river. We brace the kayaks together with overlapping paddles and break out the sandwiches and chips and bottled waters that we bought at Starbucks this morning. We eat and talk and float downstream with the seedpods, moving through splotches of sun, then shade, then sun again, spinning slowly through puffs of warm air that crinkle the river's surface. I shed my life vest and lean back in my seat. I spread my arms and tilt my face skyward in surrender to the current and the sun. I am dancing with the Passaic.

One summer a few years back my niece Catherine came to visit me on Vashon Island where I live with Kate in a house on the water. Catherine was nine that summer.

Vashon lies south and west of Seattle in Puget Sound. The only way to get there is by boat. It's a 22-minute ferry ride from downtown Seattle. The island is as big as Manhattan, with a population of about 11,000. Our house is on the north end, in the crook of a tiny inlet called Fern Cove.

One low-tide afternoon Catherine and I went for a beach walk. We picked our way along the rocky shoreline from my house west towards Peter Point, a modest jut of coastline whose steeply raked beach forms a natural levy against the Sound. The small marsh that formed behind this sea wall is a treasure trove of bleached white driftwood and polished green stones that have been pounded flat as coasters by the surf. I wanted Catherine to have some stones and wood as souvenirs of her trip.

Our progress to the Point was slowed by my habit of pausing every few steps to examine the undersides of large rocks for resident fauna. Crabs, starfish, anemones, sea cucumbers, whelks and what have you abound on these rocky Pacific Northwest beaches. Catherine was being a good sport.

We were bent over a knee-high rock whose top side wore a mop of brown bladderwort. I was pointing out a dark dripping blob of anemone. Its tentacles were retracted, and it was dangling upside down like a droopy punching bag from beneath one of the rock's slick ledges. At my urging Catherine was poking the soft anemone gently with her wet index finger.

"Why do you like nature so much?" she asked without looking up. Her question took me by surprise. I'm not sure I'd ever actually thought about why I like nature so much, although clearly I do and always have and was tickled that Catherine had taken note of this passion of mine and hopeful that she shared it.

I glanced at my niece and then out across the cold green waters of Puget Sound. Their metronomic lapping was the game show clock ticking down the seconds that remained before my answer was due. The tips of the Olympic Mountains were just visible above the hills of the nearby Kitsap peninsula. The jagged peaks were a milky blue and, even though it was mid-August, the tallest ones were still snowcapped. Time!

My answer was a platitude, but a heartfelt one. "Because," I said to Catherine, "it's so... *beautiful.*"

I've come to think better of my response to Catherine's question. The naturalist Aldo Leopold once observed that our ability to appreciate nature "begins, as in art, with the pretty," and "expands through successive stages of the beautiful to values as yet uncaptured by language." Words failed me that day, but nature never has. Its glory and wonders have always moved me. The first time I visited Seattle, I remember standing on the edge of a wooded ravine north of downtown and being brought to tears by the shafts of silver light filtering through the crowns of the Douglas fir trees. I was living in Manhattan at the time, so I was clearly starved for green. But when I think about my need for nature, how I came not just to appreciate it, but to study it and crave it, it all comes back to the lake.

My family once owned property in Andover, a smudge of a town in northwestern Jersey known for its historic gristmill and for the mines that supplied George Washington's Continental Army with iron. The family place was a pristine 100-plus-acre tract of rolling forested hills about 10 miles east of the Delaware Water Gap in Sussex County. The centerpiece of the property, and the focal point for all our activities there, was the nine-acre lake. Its official name was Auble Pond, but we never called it that. To us it was just "the lake."

Northwestern New Jersey is lake country. The Passaic River watershed extends into this part of the state, which geologists call the Highlands. It is one of New Jersey's four physiographic provinces, four stripes of land, pitched in a northeasterly direction, whose physical geography—that is, aspects of terrain, types of rock and geologic history— makes them distinct one from the other.

From northwest to southeast, these four physiographic regions are the Valley and Ridge, the Highlands, the Piedmont and the Coastal Plain. The Passaic and its tributaries flow through two of the four regions: the Highlands and the Piedmont.

Since the earth formed some 4.5 billion years ago, it has been shaped and reshaped by many cataclysmic forces. Ice sheets advanced and retreated, continents drifted and collided, sea levels rose and fell. Evidence for most of these major geologic events can be found all over New Jersey, hidden beneath the malls and subdivisions in subterranean folds and faults throughout the state. If you were to drive straight south and east from the town of Newton in Sussex County to the gambling Mecca of Atlantic City on the Jersey Shore you would pass through all four physiographic provinces and more than a billion years of geologic time.

Slicing through Sussex and Morris counties, the Highlands province is a dramatic part of New Jersey, which most out-of-staters never see and can barely conjure. The Highlands is all water and rock. Swift-flowing streams spring from its forested ridges and course through its deep valleys. The streams swell into rivers and the rivers—the Rockaway, Whippany, Wanaque, Ramapo and Pequannock—eventually join the Passaic.

Numerous lakes dot the northern half of the Highlands, a liquid legacy of the Wisconsin glacier that crept into New Jersey more than 21,000 years ago. The glacier's terminal moraine, its end line, cut the Highlands Province in half, north to south. The division is apparent on any map of the state. Dribbles of blue, like splatters on a Jackson Pollock canvas, cover the northern part of the province. The southern half looks drab by comparison.

The blue spatters are glacial lakes. Long, slender dribbles are lakes that, like Glacial Lake Passaic, were formed when glacial ice and debris clogged stream outlets. Lake Mohawk and Lake Hopatcong came about in this way. The smaller, rounder shapes—Hawthorne Lake and Lake Stockholm, for example—are kettle lakes, created when great chunks of renegade ice broke free from the retreating glacier and, stranded, began to melt, slowly, over thousands of years.

The glacial lakes of the Highlands are geologic infants compared to its rocks—enduring granites and granite-like gneisses (rhymes with "nices")—laced with marble here and there. Highland rocks are all but erosion proof, and thus as ancient as time. They were cast more than a billion years ago, deep within the earth, in a cauldron of searing heat and intense pressure that melted and twisted and deformed them. As the rocks cooled and recrystallized, they became something new. They are "the oldest rocks in New Jersey," says Otto Zapecza (pronounced Za-PET-za), a geologist with the U.S. Geological Survey in West Trenton, New Jersey. "Precambrian in age, which makes them some of the oldest rocks on earth." The Highlands is the province of memory.

The town of Andover is located in the northern half of the Highlands. The giant boulders scattered around the surrounding countryside are glacial erratics, massive rocks that were left behind by the retreating ice sheet. Our lake was not glacial in origin. A previous owner had built a small concrete dam across one of several streams on the property. A rickety wooden footbridge provided passage across the top of the dam. You had to be surefooted, because many of the bridge's cross boards were missing and the ones that remained were either cracked or loose or both.

The stream below the dam flowed into a second, smaller lake. Forgotten in the far corner of the property, this lake was under siege by the advancing marsh grass. I'm sure it's a meadow by now. Between the two lakes a spring bubbled up from below ground. It provided all the water for the house. It was hard, hard water that left a permanent crust on the bathroom tile.

The lake house was plain— red brick, one story, set on a slight rise above the lake's western shore. But it did have a few flourishes. There was a big stone fireplace in the living room, an old upright piano in the basement, and a long screened-in porch with a slate floor that ran the length of the house and overlooked the lake.

My father's father, Samuel Joseph Bruno, acquired the Andover place right after World War II. How exactly my grandfather came to own it is a mystery he would take with him to the grave. He died in 1953, the year after I was born. The tale I heard involved an Arco Welding & Machine Works client who didn't have enough cash on hand to settle his account. My grandfather asked the client what else he had. The client mentioned the property in the country. My grandfather said, "I'll take it." There are alternate stories about the acquisition of Andover. No one knows anymore which one is true. But however my grandfather came by the property one thing is certain: Andover was a great and enduring gift to us all.

My father always planned to build a second house at the lake someday, on a rise above the dam, where he and my mother would enjoy a serene retirement and where we would visit them with our own children. Sometime in the mid 1960s, the state completed another section of Interstate Route 80 West, which cut the drive time to Andover in half. This development prompted my father to suggest a kind of dry run, spending the entire summer up the lake. My mother resisted. Despite his promise to commute to and from his office in Woodridge, she suspected she'd be spending most of the summer alone in the country with five kids and a dog. And she was right. But she gave in anyway, and for the most part, I think she enjoyed those slow quiet summers in the country.

My mother was always comfortable with solitude. From time to time after my father died, we would urge her to go out more, to call her friends, to call their friends. She would smile and nod and mostly ignore us. On a few occasions she took umbrage at the implication that her life was somehow diminished by the gaps in her social calendar. Being alone, she would insist at those times, was her choice, her preference. "I enjoy my own company," she would say. There was a note of defensiveness in her comeback. Widowhood wasn't easy. It was lonely and, for a shy person like my mother, it was awkward. "I feel like the fifth wheel," she'd often say. But there was truth in her rejoinder too. My mother did enjoy her own company. I have come to appreciate what a rare and precious quality that is.

We kids passed those long summers at the lake swimming, and collecting turtles and wandering through the woods. We worked our way through summer reading lists for school. We became expert rock skippers. Some days when it was especially hot, or we were especially bored, we'd take to the porch in the late afternoon and stare out at the lake waiting for a perch or a sunfish to rise up and snatch an insect off the surface. Now and then, we'd go into Andover for ice cream, or to the drive-in theater in Newton, the big town six miles further west along Route 206. We missed our father, but we felt his presence, his quiet sheltering love.

When he was there, on weekends and most evenings, he'd play baseball with us in the yard, or give us diving lessons. My brother Joe mastered the jack knife. Now and then, he'd conscript us for a major chore, like thinning out the aquatic "weeds" that by late summer threatened to consume the lake. At least once every summer, my parents entertained. Those Andover picnics, swollen by both extended families, escalated the good times in my grandparents' backyard to an Olympian level.

From September to June we were city kids, making do with our vacant lots and zombie river. The lake was our summer camp. And it was a wonder, alive with curiosities and sensations. Pickerel with pointy snouts patrolling the whorls of *Myriophyllum*. A cottonmouth taking the sun on a flat rock in the stream below the falls. The smooth sheeting of the lake water as it spilled over the dam. Wisps of mist, like souls, that rose from the lake at sunrise. The smell of hay and dirt and dinner on muggy summer evenings. The buzz of cicadas heralding the fall. The great blue heron's squawk. The dragonflies. The lightning bugs. The way my father loved that place. The way we all did.

Carl had an Andover too, the 80-acre Arkansas farm that belonged to his paternal grandparents. Carl and his two brothers and two sisters spent their summers there. His grandfather wasn't very successful as a farmer, but he taught young Carl how to flat-pick and Carl has been a guitar player ever since. He brought his guitar along on our kayak trip so he could practice for a songwriting competition he'd entered. Carl inherited the thirties-era arch top guitar that his grandfather used to hide in the attic to keep the grandkids from messing with it. "It doesn't sound great," he says. "But you can play slide on it."

Carl grew up in Brick Town, New Jersey in the 1960s. At the time, Brick Town was a new suburban development being carved into the outer edge of the Pine Barrens, the unique, protected dwarf pine and cedar forest that covers more than a million acres in south central Jersey. This part of New Jersey lies in the Coastal Plain province, the youngest and largest of the state's four physiographic regions.

Carl's dad, James Alderson, met Carl's mother Elaine, a Jersey girl from Kenilworth, while he was stationed at McGuire Air Force Base, which abuts the Pine Barrens. James heard about the new Brick Town housing developments where homes were going for $9,000 with no money down. Who could resist a deal like that?

"Come on down, we're carving up pieces of the Pines for you," is the way Carl describes the great land rush that gave birth to Brick Town. It's a familiar Jersey tale. "It was piney woods, cedar bogs," says Carl. "Now you have a population of about 90,000. You ask people from Brick Town where the Pine Barrens is and they'll point off, 'well, it's down that way.' They don't even realize the town was in the Pine Barrens."

Carl was kind of a misfit in New Jersey. The farm, with its seemingly endless wild acres to roam and explore, was "the homeland." Brick Town was something to be more or less endured until it was time to return there.

Carl was living a double life, enthralled by the natural world in Arkansas and watching it be paved over in New Jersey. But the dissonance was enlightening. "I recognized," he says, "that we were part of the problem."

At home in Brick Town Carl and his neighborhood buddies spent their free time exploring the woods and cedar bogs on the outskirts of town. A hurricane had blown down a stand of cedars in one of the nearby bogs. The jumble of fallen trees formed a bridge between the groomed safety of the development and the deeper, darker, untamed Pine Barrens beyond. "The parents never went into the woods," says Carl. "The kids were the only ones roaming around back there, exploring the outer boundaries of the neighborhood. My mother always said, 'there's bad things going on back there.' Which only fueled our interest." One day when Carl was 16 he decided to push deeper into the Pine Barrens than he had ever been before. He tight-roped his way across the downed timber, hopped off the last cedar log and into his future.

"I got back there pretty good," he recalls. "There was this verdant swamp forest. It was a cedar bog, but with an amazing understory of skunk cabbage and fern that was such a vivid green. The canopy was just so vibrant. It was like I stepped into an otherworldly place. I was so stunned that it could be so diverse, that it had so many different layers. The vivid color of skunk cabbage; the ferns were just boundless. You know, ferns and skunk cabbage can grow at the edges, but they get tattered at a suburban margin. So I started to see, oh, the interior was different than everything we had trampled at the edge. I was blown away."

The nice thing about paddling on a river, especially when the current is going your way, is that you can make progress while doing nothing. Carl and I have probably drifted half a mile or so down the Passaic during our hour-long lunch break. But it's time to start paddling again if we want to make it to the take-out before dark.

I swipe the dried mud drippings off my laminated river map to see what lies ahead. Between New Providence, where we are now, and Chatham, where we want to be before dark, the Passaic runs in a relatively straight northeasterly direction. This is good news. The channel makes a few wide swoops coming into Chatham, swinging around a housing development and a small industrial park. But our course to the take-out will be free of the tight meanders that breed strainers and suck up time. Fueled by our chicken chipotle and turkey sandwiches, Carl and I pick up the pace.

The landscape is more suburban now, but the riverbank is still tree lined. A pair of mallards and their six ducklings waddle ashore as we pass. We scare up a great blue heron, which flaps downstream to escape us only to be startled and flushed by us again, and then again. Somewhere in New Providence, I pick up an unfamiliar scent. It's a chemical odor, the kind that leaves a taste on the back of your tongue. It's soapy and sweet with a hint of chlorine in the finish.

"Sewage outfall," says Carl, when I ask about the mystery smell. He points to a large pipe just ahead that's drizzling water into the river from a municipal sewage treatment plant nearby. Apparently we passed our first sewage outfall pipe back in Stirling, but it wasn't discharging, so I missed it. By law, all wastewater has to undergo secondary treatment before it can be emptied into the Passaic. Secondary treatment gets rid of biological contaminants such as bacteria. Primary treatment separates the solids. Tertiary treatment, or "effluent polishing," removes any remaining sediment, nutrients and microorganisms. Tertiary treatment also disinfects the wastewater, usually with chlorine.

Carl and I will pass several more sewage outfalls along the way to Chatham. Most municipal treatment plants along the Passaic comply with the secondary treatment standard. Many are even using tertiary treatment and higher. Still, I'm glad I fell in upstream of these pipes.

The Passaic and its tributaries drain 935 square miles of northeastern New Jersey and southeastern New York. This drainage area, or watershed, measures 56 miles long by 26 miles wide, and is tilted slightly northeast. On a map, it looks a bit like an egg, bounded on the west by Vernon Township, on the east by Bergen and Hudson Counties, to the south by Bridgewater Township and in the north by Woodbury Town in New York's Orange County. An estimated 2.7 million people reside in this oval basin. That's quite a bit of effluent to polish.

For the rest of today and all of tomorrow too, Carl and I will be paddling through the Central Basin, the watershed's flat, swampy, suburban, flood-prone mid-section. The Central Basin is the crescent-shaped bowl that once held Glacial Lake Passaic. Today, it is a 21,000-acre floodplain, etched by streams and home to the Hatfield Swamp and to the Troy, Black, Bog, Vly and Great Piece Meadows, the soggy leftovers of that prehistoric finger lake.

The same glacier that littered the Highlands with blue also created Glacial Lake Passaic. It was called the Wisconsin Glacier, and it formed the leading edge of the Great Laurentian ice sheet, the last ponderous mass of ice to advance as far south as New Jersey. The glaciers marched out of the north some 80,000 years ago, near the end of the Pleistocene Epoch, the Ice Age. The Pleistocene ice was vast and deep, as wide as North America and nearly two miles thick at its center. In New Jersey, says geologist Otto Zapecza, "scientists hypothesize that the ice surface was 2,000 feet above High Point, and High Point is 1,803 feet above sea level." High Point, the tallest peak in northwestern New Jersey's Kittatinny Mountain Range, is the highest point in the state. The earth's surface is still rebounding from the weight of all that frozen water and snow.

The front end of the ice inched past Paterson about 21,000 years ago. It finally ground to a halt some 1,000 years later in a wavy line that extended from present-day Perth Amboy west to Morristown. This glacial finish line is called the terminal moraine. Carl and I are some 30 miles north of the glacier's terminus now, moving through country that still bears the scars of the ice.

The Wisconsin Glacier was like a giant tanker that shoved a huge bow wave of earth before it. The rasp of boulders and gravel, sand and dirt scoured and scraped the earth, gouging great holes in some places, sealing them shut in others, and heaping mounds of debris here and there across the landscape. Basking Ridge, New Jersey was built by glacial ice.

At some point, debris from the Wisconsin glacier clogged a gap in the Watchungs near what is now Short Hills, New Jersey. This gap had offered passage to the sea for the Passaic and all the other streams that lay west of the Watchungs. With the gap now sealed shut these streams, trapped behind the Watchung ridges, backed up to form a lake.[3]

The waters of Glacial Lake Passaic soon found another outlet, a smaller gap in the Watchungs, farther south of Short Hills at a place called Moggy Hollow. But the advancing ice and its load of earth would plug that drain too.

Eventually the climate warmed, the Ice Age ended and the Wisconsin glacier began to retreat. As the ice melted, Glacial Lake Passaic grew and

---

3 The Wisconsin ice dammed rivers on the east side of the Watchungs too. Both the Hudson and Hackensack rivers existed as lakes for thousands of years. Fine clay sediments that collected on the bottom of Glacial Lake Hackensack contributed to the formation of the Meadowlands. When the ice eventually melted, sea levels rose, and the rivers flowed again. But the clay left a seal. Unable to percolate through it into the underlying soil, the water spread out across the surface of the land, transforming the old lakebed into a salt marsh.

grew and grew. At its largest, the lake measured 30 miles long and 10 miles across. Its deepest part, 240 feet down, was just south of the Great Swamp. All the towns that Carl and I are paddling past today—Millington, Gillette, Stirling, Berkeley Heights, New Providence, Summit—would have been submerged beneath tons of icy cold water.

The deepest sections of Glacial Lake Passaic survive as sodden footprints. The Great Swamp behind us, where I stood waiting for Carl, and the string of smaller marshes and meadows that lie ahead remind me that we are traveling along a very old lake bed, through a sub-basement of time. They remind me of something else too: that most lakes are ephemeral. In time, they fill in. They go dry. They disappear. They give way before a natural succession process that transforms lakes and ponds into swampy wetlands and eventually into forest. Still waters, by and large, do not last.[4] But rivers—rivers flow forever.

On Vashon Island, there's a creek that empties into Fern Cove just a little ways down the beach from my house. It's called Shinglemill Creek after the cedar shingle mill that operated along its banks in the late 19th Century. Like a patient plasterer, Shinglemill has turned Fern Cove into a wide, flat delta. Sheets of sediment fan out from the creek mouth and form smooth wide mounds of fine sand between the tendrils of rock and mussel beds. The creek cuts an ever-shifting path through this soft delta.

---

4 Graben lakes are one exception. Grabens form when a block of land sitting between two fractures, or faults, in the earth's crust slips downward, like an elevator. (Graben means "ditch" in German.) If the resulting shaft fills with water, you have a graben lake. Lake Tahoe in Nevada, Lake Baikal in Siberia and Lake Tanganyika in the African Rift Valley are all graben lakes. Lake Baikal is the deepest lake in the world; Lake Tanganyika is the second deepest; Lake Tahoe ranks tenth. The shape of a graben lake, deep and sheer-sided, makes it inhospitable to aquatic plants, which thrive in sunny shallows. With no vegetation to die and decay and accumulate on the bottom, a graben lake escapes the natural evolution that, over time, turns shallower, slope-sided lakes into woodlands.

When we first moved here in 2001, Shinglemill crossed right in front of our house and then turned sharply right to meet the Sound. After record-breaking rains in November 2007, the creek changed course completely. It abandoned the old creek bed in front of our house and carved a deeper, straighter channel from its mouth right to the surf.

Twice each day, the tide rolls in and over the new and the old creek beds. At high tide the creek disappears completely. The ducks and seagulls still know where to find it. I watch lines of goldeneye ducks and mew gulls sip and preen in the freshwater channel that is now invisible to me. Despite the twice-daily onslaught of saltwater, Shinglemill Creek doesn't falter. When the tide retreats, the creek is still there in its channel, flowing swift and strong and clear, just as before. So too the Passaic. When the glacial lake that obscured it for eons finally drained, the river was there. It had shifted course; it flowed north now instead of east. But the river was still there, flowing swift and strong, just as before.

On the way to Chatham, Carl briefs me on a minor obstacle ahead: the Chatham weir. There are actually three Chatham weirs. The one we are concerned about is a low cement wall with a notch in its center that crosses the Passaic just outside town. We have to decide whether to paddle over it or go around. It's a Goldilocks problem. If the river is too low, the wall will be exposed and we'll portage. If the river is too high, we'll portage too. Only if the river is *just right* will Carl and I paddle over the weir. Just right, according to Ed Gertler, is a matter of judgment. The weir, he writes, "might develop a strong reversal at high water, so scout."

Given the low water levels we've encountered so far today, paddling the weir seems like a distinct possibility. But Carl is a little squirrelly about the idea, because he had a close call at the weir during his 2003 trip downriver. With the flood stage Passaic running fast and deep that June, Dorina, one of the most experienced kayakers in Carl's party, got caught up in the weir.

At high water, the powerful current curls over the shallow lip of the Chatham weir and churns the river bottom on the downstream side. The agitation effectively dredges the riverbed at the base of the wall, creating a trench that can swallow a kayak and a kayaker. Dorina and her kayak disappeared into that hole.

Carl was watching from shore with the other portagers. Satisfied that there were competent rescuers at the weir, he raced down the riverbank to another take-out spot about a hundred yards downstream. If Dorina got spit out of the hole and into the current, his plan was to arrive there before she did and pluck her from the Passaic.

"I'm standing there waiting for her to bob down the river," he says. "And a frantic woman gets out of her large SUV and says, 'Oh my God, oh my God, have you seen my husband Paul? He was kayaking down the river. He's supposed to be here by now. I don't understand.'"

Carl pauses for a few seconds here to review the passage in Gertler's book that advises how and when to tackle the Chatham weir. "I want to be helpful, but I'm waiting to catch my victim of the weir," he continues. "And we didn't see any other paddlers that day. Come home three days later at the end of the trip, and there's a story on the news about nine paddlers rescued from the nearby Black River. Apparently, the Black River is fierce when there's a flood. Imagine that tiny little river becoming a torrent so strong that nine guys were swept away and had to be rescued out of there."

Though he wasn't able to offer much comfort to Paul's worried wife, Carl was relieved when he saw Dorina surface and get pulled ashore by the upstream rescue party. She suffered a gash in one knee that required stitches, but otherwise she was unhurt— and surprisingly calm given the ordeal. A less experienced paddler may not have fared so well.

With Dorina's close call fresh in our minds, Carl and I decide to table our paddle-or-portage discussion for now. It's all academic anyway until we get close enough to "scout" the Chatham weir, which is still a mile or two downstream.

Chatham is actually two distinct municipalities: Chatham Township surrounds the much smaller Chatham Borough on three sides. This is a common municipal configuration in this part of New Jersey. Together, the two Chathams cover 14.2 square miles along the eastern edge of Morris County. In terms of area and population (almost 14,000), Chatham is about half as big as Basking Ridge. Demographically, the two Passaic River towns are nearly identical: mostly white folks whose median household incomes are more than twice the national average.[5] Chatham's slightly wealthier residents are just less densely packed— and within easier commuting distance from Manhattan.

The terrain gets hillier on the outskirts of Chatham. The Passaic lopes south and east around a housing development and slips under the bridge at Mount Vernon Avenue. Once past the bridge, the river begins a gentle downhill run into Chatham. It winds through a narrow green belt, then executes a wide U-shaped hook to the northwest, wrapping itself around a small industrial park like a bracket. Carl and I glimpse the backs of factories and warehouses on the banks above the river. The channel is straight now and studded with pale boulders the size of dishwashers. The current is strong. It's a slalom course. My tired arms are grateful for the boost. Oh to be swept into Chatham!

---

5 Source: U.S. Census Bureau 2005-2009 American Community Survey.

Alas, low water denies us once again. The river is actually deeper here than it was through the Millington Gorge—knee-, as opposed to mid-calf-deep. But the rocks on the riverbed are bigger too, so the result is the same. Carl and I reprise our Millington Gorge routine several times along this stretch of the Passaic, but with a new twist. Without the ballast of our bodies pinning the boats to the rocky bottom, the Pungos become like wild stallions every time we step out. They buck and spin and tip and rock in the swift current and generally try to escape downstream. It's hard to steady the Pungo long enough to lower myself back into the cockpit. I finally decide to straddle the cockpit, then quickly plop down into my seat before the boat can scoot out from under me. It's not a bad strategy, but I'm not always quick enough. Now and then my butt lands on the stern, just aft of the cockpit, and I have to start all over again.

I glance over at Carl. He's struggling too.

I realize that my body type— short and compact compared to Carl's long and lanky— confers certain advantages in these river situations. It's much easier for me to swing my short legs in and out of the boat, and since I'm lighter, my Pungo doesn't get hung up on the rocks quite as often. My shortness will be a handicap when it comes time to hoist the kayaks back onto the roof of the car. But right now, in this riffle, morphology is working for me.

The river gentles at the end of this whitewater sluice. We enter a quiet shady stretch. We paddle under a beautiful old arched stone railroad bridge. I see the weir ahead, a faint straight line etching the water's surface. Both ends of the wall are exposed, but there's a wide, steady stream of river passing over the center. Carl paddles closer and scans the weir. I can tell he's tempted. In the end though, we play it safe, forsaking our vow and portaging the boats. It's an easy maneuver. We wedge the Pungos between the wall and the flatter, drier right bank and step out for what I hope will be the penultimate time today. There's a park on this side of the river, just off through a shallow screen of trees. A family is having a picnic. Dad is flipping burgers on a grill. They smell divine.

The sun is going down now. There's probably an hour of daylight left. The air is starting to feel cooler, almost chilly. We're probably 300 yards from the take-out and for the first time today I'm eager to get there. It's been a long and amazing day. All my pre-paddle fears—that my arms would fail me, that Carl and I would hate each other, that I'd fall in—have proven groundless or bearable. Despite the strainers and the trash and the low water and the occasional sewage outfall, today's paddle on the upper Passaic has been kind of wonderful, like a blind date that turned out fun. I still have my guard up. I know there will be ugliness ahead. The DSAYs will start piling up once we hit Paterson. But right here, right now, at the tail end of my first real encounter with the Passaic, it seems possible that the river isn't such a monster after all. And this realization stirs a memory of my mother.

She is coming slowly down the stairs from the master bedroom where she's been "getting ready" into the kitchen where we kids are awaiting her entrance. My parents rarely went anywhere without us, but on this night they are attending some dress-up affair, probably a Rotary Club dinner. (My father was a member.) My mother is wearing a midnight blue cocktail dress and black high heels. The dress is sleeveless with a scoop neckline that shows off her décolletage. Her heels have black straps that wrap around her ankles. Her brown hair is curled, her makeup perfect.

"Mommy," gasps my sister, who at age four was already a fashionista, "you look gramulous!" My mother bursts out laughing and hugs my sister against the blue dress. Cathy meant "glamorous," of course. And she's right. Our mother looks like a movie star.

My mother hated getting old. She stayed remarkably healthy right up until the last two years of her life. But every little loss, every failing became a source of great frustration, shrinking her world and wounding her pride and threatening the independence she held so dear. For my mother, getting old was like living in a cage that kept getting smaller and smaller. Her eyes got worse, so she gave up driving at night. Bunions on both feet, badges from her nursing years, kept her walks short and made it nearly impossible to find shoes that fit. Moles on her back spelled the end of bathing suits. Forgetfulness was the worst thing. "Don't get old," she'd tell us.

Family members and friends were always dropping in on my mother. My friends and my siblings' friends all loved her. She was good company. She made everyone feel welcome, sitting across from them at the kitchen table, sipping black coffee from a cup and saucer—never a mug—and listening. She loved to listen.

I think old age was so hard for my mother because she had to go through so much of it alone. Solitude is easier to savor when the alternative is close at hand. By the time my mother reached 73, most of her dearest contemporaries were dead—my father, her parents, her only sister Dorothy, and all but one of the cozy gaggle of childhood pals she referred to as her "club girls." My mother had no one her own age to commiserate or hang out with. Worst of all, there was hardly anyone left who could remember the vital young woman she had been. The glamour girl in the blue cocktail dress; the strong, smart, independent, capable nurse; the wife, daughter, sister, young mother. They all vanished. My mother, in her old age, was reduced to a shard of her magnificent, multifaceted self. I imagine that people who only knew her in the final years of her life must have seen a pleasant old lady who had a tendency to repeat herself and a hard time keeping the names of her grandchildren straight. Nothing more.

Growing up along the lower Passaic is like meeting my mother in her dotage. It is never seeing or imagining the beautiful, vibrant river that used to be. Today, I got a glimpse of that river, and like my mother it is lovely.

Visions of my mother and a pre-industrial Passaic, and a hot shower, dry clothes and a cold beer crowd my thoughts as I paddle on downstream towards Chatham, which might explain— in part, at least— how I missed the take-out.

Carl and I are aiming for Chatham's Sheppard Kollock Park[6], a pocket of green with a baseball diamond at one end and a small concrete boat ramp at the other. The ramp is about 100 yards from Henderson Street, the short, shady block where we parked my rental car last night. Before we left the park last night, Carl walked the river. He pointed out a few landmarks that would signal the approach of the park and the boat ramp. I thought for sure I'd recognize the take-out from the water. How can you miss a boat ramp? But things look different from the river. Your perspective skews. Your sense of time and distance— all shot to hell. The familiar becomes unfamiliar, even inscrutable. It's like being stoned.

I get ahead of Carl as we near the park. There's a young fisherman standing in the middle of the channel. He is wearing green hip waders and one of those short khaki angler vests with all the pockets. He is just downstream of the boat ramp, but I don't notice the ramp, because I'm focused on the fisherman. He has cast upstream in my direction. I'm calculating the best way to avoid his line, which is now starting to slacken and drift

6 Born in Delaware in 1750, Sheppard Kollock was an editor and a printer who published several books, including *Ebenezer Elmer, Surgeon of the Regiment*, before becoming a First Lieutenant in the Continental Army. When General George Washington decided that a newspaper would help promote the rebel cause, he asked Kollock to spearhead the effort. Kollock published the first edition of the *New Jersey Journal*, from Chatham, on February 6, 1779. He died 60 years later, in Philadelphia, at age 89. His paper, renamed the *Elizabeth Daily Journal*, is still coming out.

downstream in the current. When I get within 30 yards of the fisherman, he gets a bite. He yanks hard on his pole to set the hook. I watch him reel in a fish that looks to be at least a foot long. He holds his wriggling catch up over his head and, grinning, yells out, "Thank you!" He assumes I herded the fish his way and, hell, he may be right. I am still graciously acknowledging his gratitude when I speed past the boat ramp. *Shit!*

I dig my paddle into the water as hard as I can, and bring the Pungo around. But there's no way I can paddle against this current, so I just, sort of, bail out. Elegant? Hardly. Effective? Very. I walk my Pungo upstream to the exit ramp where Carl and dinner and a dry pair of pants await. Terra firma has never felt this good.

6 | GREAT PIECE MEADOWS

AT BREAKFAST THIS MORNING, Carl proposes a change in today's paddle plan. Given the river's sluggish pace, and the likelihood of a strainer or two along the way, he's doubtful we'll be able to cover the whole 27-mile distance before dark. He suggests skipping the initial stretch between Chatham and West Caldwell. This will mean missing the point at which the Rockaway River joins the Passaic. But it will give us more time to savor Great Piece Meadows, which according to Carl is the wilder, prettier part of the paddle. There's one other upside to this suggestion, which Carl doesn't know about: my nephew Joseph is making his First Holy Communion today. Joseph is my brother Joe's youngest child, and my godchild, which means that, technically, I'm responsible for his spiritual development. There's no way I'll make it to the church, but with this shorter route I may be able to get to Joe's house in Westfield, New Jersey before the Communion party ends. Okay, I tell Carl. I'm in. Now all we need to do is find a new put-in.

We finish breakfast, pack the cars and strike out for Camp Lane in Fairfield, today's take-out location. Fairfield is nearly 30 miles north and east of our Basking Ridge hotel, and after we stash my rental Ford there, we still have to locate a new launch point. But it's another warm, sunny day—we've been lucky with the weather—and with our newly abbreviated itinerary, there's no need to rush. I'm settling in for a leisurely Sunday paddle.

I follow Carl up Route 287 North to Route 46 East. We take the Passaic Avenue exit and bump along through a small industrial park, one of many that crowd the highway. Once we pass through this industrial portal, we hang a right and I find myself on a green, tranquil country road that quickly dead-ends at the river's edge. This is Camp Lane.

"It's pretty great, isn't it?" says Carl, when we step out of the cars. He's enjoying my amazement.

Carl discovered Camp Lane while he was scouring maps in search of prospective take-out spots for his 2003 Passaic excursion. He was looking for any road that got close to the river. Camp Lane fit the profile perfectly. As its short paved part nears the Passaic, Camp Lane just devolves into a wide gravel path that slips into the water. We drive through a copse of trees and stop in a small, graveled pullout at the end of the road. Across the street from this parking area is a two-story, patchwork dwelling, not 30 feet from the water's edge. An American flag flaps gently from a post on the front porch. Plywood covers two large first floor windows. There's a blue pickup in the driveway and a white motorboat and four small tractors lined up in the front yard.

"That's old John's house," says Carl. Old John offered to let Carl and his 2003 paddling group camp in his front yard on their second night out. Unfortunately, heavy rains scuttled the camping plan.

"We got to John's house around 6pm and he was waiting, as promised," recalls Carl. "His house was more boat than house at that point. He came out in a pair of waders, because his first floor was underwater. His wife was up in the second floor window eating a bowl of cornflakes. She was pretty nonchalant. When she finished eating she just popped open the window and poured the extra milk and the last few cornflakes out into the river."

Old John helped tie the 20 kayaks up to a telephone pole in his yard. He took all the life jackets and paddles inside the house and hung them up as high as he could get them. When everything was secure, Carl and the kayakers left to spend the night in a nearby hotel. Old John went back upstairs to join his wife in their still dry bedroom. They'd been through soggy nights like this before.

The front door to Old John's place is ajar this morning. Carl calls out a few hellos but no one seems to be home. We take a quick stroll downstream along a footpath that parallels the Passaic. The river is narrow here, about 100 feet across, and it bows slightly, wrapping Camp Lane in a gentle embrace. Mature oaks shade the footpath. The opposite bank is wooded too, but I can see a sunny clearing beyond the trees. On our map a few small roads etch that side of the river. From here, there's not a human dwelling in sight.

I once visited an old college roommate in New Orleans, her hometown. We spent a day poking around bayous in a leaky 14-foot skiff. The Passaic River at Camp Lane reminds me of those slow-moving Louisiana backwaters—a hidden, beautiful secret. I can see why Old John and his wife were loath to leave this spot even for a night, even with their ground floor submerged. This thread of road in the far northeast corner of Great Piece Meadows is a gem of a boat launch. It will serve as the endpoint for today's paddle, and as the starting point for our next weekend river trip, five months from now in October.

Carl and I leave the quiet of Camp Lane reluctantly. We climb into the Camry and get back on Route 46, heading west this time, on the lookout for a manageable put-in.

Route 46 is a brutal shock after Camp Lane. The road is four lanes wide with low concrete Jersey barriers separating east and westbound traffic. Both sides of the highway are thick with strip mall businesses, car dealerships, multiplexes, big chain stores, big malls, small diners, light industry and the acres of asphalt parking lots that support them. We plan to exit near Bloomfield Avenue in West Caldwell, just before the highway passes over the Passaic. Carl is convinced we'll find a boat launch somewhere below the Route 46 Bridge. I can tell he's winging it. But I'm not worried. Carl has a gift for improv.

The Passaic River cuts an uncharacteristically straight path from Chatham north to Great Piece Meadows. The 15-mile stretch of river below us, as well as the 10 miles we will actually paddle today flows through the very heart of the Central Basin, the long elliptical depression that once brimmed with the icy waters of Glacial Lake Passaic. Reminders of the Central Basin's watery beginnings will be everywhere along our route today. More than 30,000 Central Basin acres, about 34 square miles, are wetland: the Great Swamp, Hatfield Swamp, Black Meadows, Troy Meadows, Lee Meadows, Great Piece Meadows, Little Piece Meadows, Bog Meadows and Vly Meadows. Taken together, they constitute one of the largest wetland complexes in the state. With the exception of Great Swamp, which is already behind us, Carl and I will kayak past or through all of these old lakebed hollows today as we make our way back to Camp Lane.

Carl veers right off Route 46 into a huge, mostly deserted parking lot. We proceed past a tiny white box, the Cutting Board Bar & Grill, and make our way to the riverbank where we park the Camry beneath a giant billboard advertising stainless steel Viking grills from Karl's Appliance (outlets in Fairfield, Madison, Orange, Perth Amboy, Newton, Belleville and

Montclair). The Route 46 Bridge is just upstream to our left. The Passaic is directly in front of us, across a wide flat lawn that recalls the old Homelite field. The lawn ends in a sheer-sided dirt bank that plunges down to the river. There must be a better place to put the boats in the water. "No, this is going to work," says Carl, sensing my crisis of faith.

He strides over to a group of five people who are fishing and picnicking on the grass. Three middle-aged men slump silently in lawn chairs, their poles propped on the Y-shaped branches they have planted in the soft grass. Two women are busy arranging containers of food on the long table they've set up in the shade of a nearby warehouse. They are chatting away in Russian or Polish or some Eastern European dialect.

The U.S. Fish and Wildlife Service lists 35 different species of fish in the central Passaic River Basin. The New Jersey Department of Environmental Protection says it's okay to catch and eat them all, albeit in limited quantities. Dioxin is not a threat here in the upper Passaic. Like the once plentiful runs of shad, dioxin is prevented from migrating this far upriver by the Dundee Dam, which is some 30 miles downstream in Garfield. Still, fish caught in this stretch of Passaic may contain dangerously high levels of PCBs. The DEP cautions the average healthy person to limit his or her consumption to one meal per week. On its "Fish Smart Eat Smart NJ" web site, the DEP even provides tips— with diagrams— on how to safely prepare your Passaic River catch:

"Do not eat the heads, guts or liver, because PCBs usually concentrate in those body parts. Also, avoid consumption of any reproductive parts such as eggs or roe.

"Use a cooking method such as baking, broiling, frying, grilling, or steaming that allows the fats and juices to drain away from the fish.

"Avoid batter, breading or coatings that can hold in the juices that may contain contaminants. The juices should be thrown away since they contain the PCBs and other chemicals that were in the fat. Do not pour these juices over the fish as a sauce or to moisten the fish.

"After cooking, discard all liquids and frying oils. Do not reuse."

I hope the members of this fishing party have Internet access.

With gestures and smiles Carl conveys our need for a better boat launch. One of the women, the sturdy one with short bleached blonde hair and big green eyes, is quick to grasp the situation. She leads us about 50 yards downstream to where a small elderly man is casting from a more gently sloping ledge. This is more like it. We thank her and set to work unloading the boats. Just as we are about to shove off, the old fisherman lands a big one. It might be a trout, but before we can make a positive ID, he bundles his catch into a white plastic grocery bag and hurries off without a word.

Within minutes of launching the Pungos, Carl and I enter Great Piece Meadows. The Route 80 Bridge is up ahead, about a half mile downstream. Already we can hear the distant hum of highway traffic, punctuated by the occasional pop-pop of gunfire from the nearby North Jersey Gun Club in Fairfield and the drone of small aircraft on approach to the Morristown Municipal Airport. The airstrip was built on top of Black Meadows.

The traffic din peaks as we close in on Route 80. The Interstate extends from the western shore of the Hudson River all the way to San Francisco. We paddle under the Route 80 Bridge, which is thick with 18-wheelers at the very beginning or the very end of their cross-country runs. The racket fades with every paddle stroke thereafter. Soon, there are only river sounds—the rhythmic plunk of paddles, the shushing of the trees, the trills and whistles of songbirds. The river ahead is wide. It winds lazily through a sun-drenched landscape of new spring green.

For the first few miles the Passaic clings to the western edge of Great Piece Meadows. Here and there, the river passes close to neighborhoods on the eastern edge of Montville, but the people and their houses remain hidden behind a thick screen of trees. Carl and I are alone again on a new reach of the Passaic, which is placid and empty and very, very lovely.

I thought I knew my home state pretty well, at least the northern half of it where I grew up. But somehow I missed this large, lush wilderness swamp just 12 miles from North Arlington.

Great Piece Meadows is the northernmost tip of the Passaic's Central Basin. At roughly half the size of Great Swamp, Great Piece is the Central Basin's second largest wetland. The "meadow" in its name is a little misleading. Portions of Great Piece are indeed marshy grassland, which prompted this description from poet John Alleyne MacNab in his 1890 book *Song of the Passaic:*

And vast expanse of "Waste Lands" fill
With ooze of water, at its will,
And marshes, sodden, low and dank,
O'ergrown with grasses, wild and rank.

But today trees and vines dominate the Great Piece landscape. Many resident species, including the river birch (*Betula nigra*), sweetgum (*Liquidambar styraciflua*) and American sycamore (*Platanus occidentalis*) trees and the vine known as Virginia creeper (*Parthenocissus quinquefolia*) are classic floodplain foliage. Edward Gertler, in his 2002 paddling book, calls Great Piece Meadows a "swamp forest," which these days is a more accurate description.

The Meadows anchor a tight cluster of five fragile wetlands that buffer the Passaic about 15 miles downstream, or north, of Chatham. The flood-prone cluster includes the smaller, conjoined Bog and Vly Meadows, which lie due north of Great Piece, and the larger Troy Meadows and Hatfield Swamp to the south. On a map, these five blobs of green form an armless stick man; Bog and Vly Meadows are his head, Great Piece his ample torso, and Troy Meadows and Hatfield Swamp his stubby right and left legs, respectively. The stickman represents nearly 10,000 empty acres, a rare safe harbor for wildlife and migratory waterfowl just 25 miles west of Manhattan.

Surrounding the stickman (clockwise from the southwest) are the bottomland towns of Whippany, Troy Hills, Parsippany, Montville, Lincoln Park, Fairfield, West Caldwell, Roseland and East Hanover. Neighborhoods in some of these towns— Lincoln Park and Fairfield, for instance— are built on drained marshland. Development reigns to the south and east of the Central Basin lowlands. But the population thins as you move north and west into the still semi-rural reaches of northwestern Morris County and southeastern Sussex County. The stickman represents an ecotope on the edge, a patch of natural world poised precariously between what is and what used to be.

Thirty six different species of mammals prowl this patch: minks, coyotes, beavers, foxes (both red and grey) and river otters. More than 220 bird species rely on the meadows as either a stopover during their grueling spring and fall migrations or, in the case of the 90 nesting species, as a permanent home. Forty four different kinds of reptiles and amphibians live here. The meadows are the only place in New Jersey where you can find a blue-spotted salamander (*Ambystoma laterale*) or a bog turtle (*Clemmys muhlenbergii*). Both top New Jersey's endangered species list. Other endangered or threatened animals and plants that find refuge in these Central Basin wetlands include bobcats (*Lynx rufus*), barred (*Strix varia*) and long-eared (*Asio otus*) owls, wood turtles (*Clemmys insculpta*), the buttonbush dodder (*Cuscuta cephalanthi*) and the Louisiana sedge (*Carex louisianica*). The area doesn't enjoy the kind of federal protection that guarantees the survival of the Great Swamp National Wildlife Refuge to the south, which is ironic since Great Piece and its neighboring meadows and swamps play a far more critical role in containing Passaic River floodwaters. During rainy spells, the river regularly escapes its banks and creeps out into these forested grasslands.

The Passaic enters this five-wetland complex from the south. It passes well east of Troy Meadows and flows straight through the Hatfield Swamp before entering Great Piece Meadows. The Passaic drains, but never touches the northernmost Bog and Vly Meadows. The river remains entirely within Great Piece Meadows until it leaves this whole swampy section of its watershed just above Two Bridges, where it welcomes the Pompton, its largest tributary.

Great Piece Meadows is a triangle with sides that droop like a pout. The triangle straddles two towns and two counties: Lincoln Park in Morris County to the northwest and Fairfield in Essex County to the southeast. A 10-mile stretch of Interstate Route 80 cuts through the southeastern edge of Great Piece Meadows. The highway and the river intersect in two places, at the extreme southern and northern corners of the wetland. Except for these two encounters, the river and the highway keep a safe distance from each other, like boxers in their corners. The Passaic wiggles north through Great Piece Meadows; the shallow convex arc of I-80 frowns along the meadows' southern edge.

The highway hews to the rule that the shortest distance between two points is a straight line; or, in this case, a close approximation of a straight line. But the river has no interest in going straight. The Passaic proceeds generally north through Great Piece Meadows, but its path is wildly indirect, convoluted, intestinal. This is not the menacing river of my youth, imprisoned in its skanky trench. This river is free. It carves an impulsive channel, lurching this way and that and folding back upon itself, sniffing every nook and corner. When the Passaic enters Great Piece Meadows it becomes a hound off leash.

Satellite images of Great Piece Meadows reveal the Passaic's looping present-day course, and its many old abandoned cuts. The dried up detours stand apart and disconnected from the main channel, some at considerable distances. Isolated meanders like these probably existed for a time as oxbow[1] lakes, a type of lake that only forms in floodplains.

Oxbows begin as pronounced parabolic bends in a river. Mounting sediment eventually pinches oxbows off from the main stream. When that occurs, they stand alone as lakes. Once riverine, they become lacustrine, or lake-like. Oxbow lakes retain the curvature of the bend that formed them. Some stay narrow like garden hoses. Others widen into boomerangs or quarter moons. From overhead, you can still make out the imprints of bygone oxbows in Great Piece Meadows. They linger on as faint hooplike patterns in the vegetation, shadow lakes, palimpsests.

I like to think of the river's meanderings through Great Piece Meadows not as confused or aimless, but as curious, unhurried, deliberate explorations. Each loop is a scouting party sent off by the river to recon an unknown edge or corner of its floodplain frontier. Each oxbow lake is a colony left behind by the river, a loop of its own river DNA that will give rise to a lake, a new and wholly different type of ecosystem.

This is the nature of a river, the way it is born to behave, or misbehave; to flow ever downstream, scouring outside banks, building inside banks, wandering back and forth across its floodplain, moving in more than one direction at a time. In *Water: A Primer*, the late Luna B. Leopold, hydrologist and son of famed naturalist Aldo Leopold, wrote that "the progressive sideways movement of the [river] channel is to the left in one

---

1 The name oxbow comes from the U-shaped collar that secured the wooden yoke around the necks of oxen teams.

place and to the right in another. Thus, given sufficient time, the channel will eventually occupy every position within the valley." How much time is "sufficient" will depend upon the power of a river's current, the composition of its valley floor and the brute vagaries of nature. The Passaic's curiosity was stalled for eons by oppressive glacial ice. Still, the ghost oxbows in Great Piece Meadows stand as evidence of the phenomenon that Leopold describes, proof of the hound at work. They are the Passaic proclaiming: "I was here, and here, and here…"

Through most of yesterday's paddle, the Passaic felt like a creek, narrow, shallow, shaded and small. Here in Great Piece Meadows, swollen by the arrival of the Rockaway, which came in a few miles upstream, the Passaic is finally, thoroughly a river. The channel is wide and deep. Tall trees line the banks—oak, maple, shagbark hickory. Their lower branches brush the water's surface. We encounter a handful of strainers. A few are big enough to block the entire channel, and we jam-o-ram through them handily. Only one strainer presents a challenge.

We've been paddling for a couple of hours when we come across it, the widest strainer we've seen so far. There are about a half dozen downed trees collected at this one fairly narrow point in the river. Their trunks span the channel in parallel to one another, creating a kind of floating log road. There are a few good-sized tunnels through the hive of branches, but how best to poke the boats through the gaps is the subject of some discussion. The trunks are too close together to deploy the Up 'n Over or the Teeter-Totter methods. With so many trunks to traverse, the Ram-o-jam seems impractical, but we default to it anyway.

With a strong paddling start Carl punches my boat deep into the strainer's widest hole. I slide my paddle inside the cockpit and pull my Pungo through to the other side branch by branch. Carl's boat gets hung up inside the strainer. He is forced to step out and cross the final few logs on foot, pulling his Pungo along behind him. I'm amazed when he makes it to the end without falling in. The man has a future in logrolling. But the challenge isn't over when Carl gets to the last log. He still has to get back into his boat. The Up 'n Over reentry won't work here because there isn't enough room inside the strainer to pull the Pungo parallel to that last trunk. Carl has to bust a new move: the Roy Rogers.

He points the Pungo downstream, steadies it for a second, then hurls himself on top of the slowly retreating kayak. He lands with his legs astride the stern, like Roy straddling Trigger. Then he shimmies up and into the cockpit. It is a bold, if less than graceful move, and I have to marvel at Carl's fearlessness when it comes to the Passaic. Getting soaked with river water or covered in river mud doesn't seem to faze him. He has none of my squeamishness about the Passaic. Or if he does, he keeps it well hidden.

All morning long, Carl and I paddle side-by-side, chatting and pointing out sites to one another: the little inlet abloom with yellow spatterdock flowers, the pair of donkeys inexplicably ambling through the woods. We let the kayaks drift over to the banks now and then to get a closer look at some submerged aquatic plant that is growing in the shallows. At one sharp bend in the river, we spy a small flock of sheep grazing on a wide green hillside that slopes gently down through a grove of trees to the water's edge. Passaic River pastoral.

Just before lunch, Carl and I pause in the middle of the channel to take a reading with the Secchi[2] disk that Carl has brought along. A Secchi (rhymes with Trekky) disk is a flat white, or black and white circle, about

---

2 The Secchi disk was invented in 1865 by Father Pietro Angelo Secchi, a Jesuit priest who ran the Vatican Observatory. Secchi is best known as an astronomer. Craters on the moon and on Mars are named for him. He developed the Secchi disk for the head of the Papal Navy, who had asked Secchi for a way to measure the clarity of the Mediterranean Sea.

a foot in diameter, with a string attached. It is used to measure water clarity. The string of a Secchi disk is graduated in inches. To gauge transparency, you lower the disk into a river, or lake or ocean, and note the depth at which the disk disappears. That tells you how far light is penetrating down into the water, which is a measure of water clarity or transparency. Of course, transparency is a mutable condition. Wind and current can stir up sediments and uproot plants, which muddy the waters. Blooms of algae can limit transparency too. Carl's Secchi disk disappears after about two feet. Suspended sediment and plant matter here in Great Piece Meadows make the Passaic murkier than it was in the Great Swamp yesterday where we logged a Secchi reading of nearly four feet.

The clearest water I have ever seen was in South Carolina. I took a job there in 1979, a few months after I finished my master's degree. I was the algae specialist on an 11-member research team that was studying polycyclic aromatic hydrocarbons. PAHs, as they are known, are carcinogens, similar to dioxins in that both are by-products of incineration. PAHs get released whenever something organic is burned. So, for example, forest fires and backyard grilling release PAHs into the atmosphere. With a grant from the U.S. Environmental Protection Agency, our team was designing a computer model to predict where these airborne PAH compounds eventually settled once they got into stream systems. How much, for instance, wound up in the water, how much in the sediments, how much in the stream's plant and animal communities?

We were doing the research at the Savannah River Ecology Lab. It was a University of Georgia field station located on the site of the Savannah River Plant, one of the largest nuclear facilities in the country. The Savannah River Plant is in Williston, South Carolina, but most of the people who worked there lived about 10 miles west in Aiken, the westernmost town in South Carolina. Aiken was a retreat of sorts, first for wealthy Charlestonians eager to escape the summer heat, and then for wealthy Yankees eager to play the ponies. The town was the winter training ground for many of the thoroughbreds that raced up north at Belmont and Saratoga. It was also one of the most beautiful towns I've ever lived in. Ancient live oaks tunneled its broad avenues; azaleas blazed in every yard; magnolia and tea olive and jasmine blossoms perfumed the air. I used to ride my bike a lot in Aiken. I could literally smell my way home. It was the first and only time I have ever lived in the South, and I loved it.

I chose a stream called Upper Three Runs Creek for my algae research. Upper Three Runs was a minor tributary of the Savannah River. It was a swift, sinuous black water stream, which had the great good fortune of flowing through a landscape that was largely devoid of humans. The Savannah River Plant was a 300-square-mile tract in the Sand Hills of South Carolina, a sparsely populated region in a sparsely populated state. The federal government bought up five small towns to make room for the nuclear facility, which began making plutonium in 1956. There are surely downsides to flowing through a nuclear reservation, but if Upper Three Runs Creek had been a minor tributary of the Passaic—the First River in Belleville, say—it would have been polluted or diverted into some subterranean culvert to make way for houses, highways, rail yards or shopping malls. Either way, its magic and power would have been lost.

Tannins from surrounding trees and shrubs, like the tannins in Bazil's Rio Negro, turned Upper Three Runs Creek the color of tea. But the water was so clear that even in the creek's deepest pools you could make out every ripple in its white sandy bottom. Peering down into the creek was like looking through tinted glass. Secchi disk readings of eight, or even 10 feet were common. The only thing preventing higher Secchi readings was the creek bottom itself.

In addition to staining the water, the tannins in Upper Three Runs also lowered the pH, which made the creek a perfect habitat for acid-loving diatoms, a type of algae whose cell walls are impregnated with silicon, literally made of glass. These microscopic greenhouses were my specialty. The first time I saw one I was smitten.

The most common diatoms in my Upper Three Runs' samples belonged to the genus *Fragellaria*, a needle-shaped species that creates long flat chains the color of single malt scotch. Other regulars at my study site included a nasty black water snake, a chattery pair of nesting yellow warblers and Mike Allred, a post-doctoral candidate at the lab. Mike was studying leaf litter; or, more precisely, the rate at which the invertebrates in Upper Three Runs broke down the leaves, twigs and other organic matter that fell into the water. Mike was tall and lanky like Carl with fine, straight brown hair that was often in his eyes. Mike was an Atlanta boy. He spoke in a refined, city drawl that was more honey than twang.

Mike designed and built four floating platforms to collect his data. They were about as long as jet skis and looked like pallets on pontoons. Each had a small, submerged chamber where Mike stuffed mesh bags filled with precise amounts of leaf litter. To hold the platforms in place against the current, he rigged a spider web of white rope that crisscrossed the stream, connecting the platforms to each other and to the two loblolly pines that anchored the whole apparatus from opposite banks. I kept my diatom collectors on Mike's platforms. The collectors were 18-inch-long Plexiglas planks with 20 grooves on their surfaces that held glass microscope slides. I secured the slides with fishing line. In a week's time, creek diatoms would colonize the smooth glass surfaces.

Mike and I drove out to the site together at least one morning a week to collect samples and check on things. We'd get to the lab early on those mornings so we didn't get stuck driving 1429, the biggest, oldest, orneriest pickup in the lab's 10-vehicle fleet. It was a 20-minute ride from the lab to Upper Three Runs along narrow sandy roads that meandered as much as the creek itself.

When we arrived, we'd launch the aluminum dinghy we kept at the site and pull ourselves along the rope lines from platform to platform. The sight of Mike's 6'7" frame hauling us around in that tiny rowboat was comical and nerve-wracking. His center of gravity was dangerously high, and the black snake had a habit of striking the boat. But in the six months we spent sampling at Upper Three Runs, Mike never faltered and neither did I. The only mishap we ever had was the morning we left the truck windows rolled down and returned to find the cab full of biting deer flies.

Being in Great Piece Meadows with Carl right now brings back those mornings on Upper Three Runs Creek with Mike Allred. It has to do with the water and the boats and the quiet and the high I always get from being away from the "office" on some outdoor adventure. But it's something else too. It's the kinship of a fellow nature lover, in whose company you can be comfortably alone together in the wild.

From time to time in the years after my father's death, my mother would talk about selling Andover. We knew she was serious. She needed the proceeds from the sale to make ends meet. But we chose to remain in denial, cocooned by a sentimental certainty that no matter what my mother said, she could never actually let Andover go. In the spring of 1980, while I was living in South Carolina, she proved us wrong. My mother sold the lake.

My brothers and sister and I begged her to reconsider. Paul and Joe were still in school at the time, but Johnny, Cathy and I had real jobs. Johnny was a Marine lieutenant stationed in Okinawa, Japan. Cathy was a Registered Nurse like my mother, working in the recovery room at Holy Name Hospital in Teaneck. I was the diatom specialist in Aiken. We offered to help with the cost of insurance, which along with property values had escalated with the opening of ski resorts at nearby Great Gorge and Vernon Valley.

Andover property values seemed to inch up with the completion of each new leg of Route 80 in the 1960s. The final stretch of Interstate opened up rural northwestern New Jersey the way the road to Paulus Hook had opened up 18th Century Newark. With the arrival of the ski industry in the 1970s, sleepy Sussex County became a winter destination, and the land grab was on. The area's farmers, as financially strapped as my mother, began selling off parcels of their own now valuable land. Every sale pushed property values and taxes higher, which only forced or tempted more sales and more development. And that's the way it often happens: a financial imperative, some gentle arm-twisting, a quiet sale, maybe a little back-room wheeling and dealing to sooth a critic or sidestep a zoning law. The changes aren't alarming at first. "Will you look at that, honey, old man Riggins sold off his lower pasture." But each transaction is another step down the path to cataclysmic change. By the time you realize what's happening, it's already too late. Construction crews are digging up the old pasture and pouring foundations for the 200-home development that will rise in its place. Pretty soon the whine of the weed whacker supplants the song of the meadowlark.

It seems petty to begrudge the new neighbors their version of the American Dream, or the former pasture owner his financial security, or the town fathers their heftier tax base. But if you had ever walked by that old pasture of a summer morning, with the dew-drenched grass bowing to the coming sun and the smells of hay and earth saturating the soft moist air, you will find it hard to ignore the feeling you get as you drive past the rows of new single-family houses. That heaviness in your chest is the burden of what's been lost, and it never goes away. Aldo Leopold gave voice to this ache in *A Sand County Almanac* when he rued the loss of the passenger pigeon (*Ectopistes migratorius*). The great flocks were hunted to extinction in the 19th Century. "The gadgets of industry bring us more comfort than the pigeons did," wrote Leopold. "But do they add as much to the glory of the spring?"

With her share of the Andover sale my mother paid college tuitions for Joe and Paul, replaced the roof on our North Arlington home and opened a modest retirement account. We all knew that we'd eventually lose the lake, that the sale was inevitable, that my mother really didn't have much choice. But it took us a while to forgive her, and I am ashamed for that.

Carl and I stop for lunch at a spot in the river where the gnarled knee of an old oak tree genuflects down to the river's surface. We bring the kayaks alongside the trunk, and step along its knobby surface to shore. After beaching the Pungos, we set out on foot to find a good picnic spot, someplace sunny and free of poison ivy, which makes a minefield of this river valley.

The terrain here in Great Piece Meadows is flat and open. The ground is dry and firm underfoot and covered with low blonde grass and a mix of hardwood trees whose dead trunks and downed branches invite perching. The wide clearings between the trees are bowls of sun. Several small swales bear the scars of all-terrain vehicles. The tire treads have ripped through the dun-colored duff, exposing a thick dark mud below. There is a light film over everything, a thin crust of river sediment left behind by the annual spring soaking. The whole place could use a good dusting. The river has been here all right, and it has been here often.

We pick a medium-sized trunk near the edge of a small, sunny clearing for our picnic. It's about 50 yards from the boats. Before we commit we walk the perimeter on a poison ivy patrol. Not too much. We're good to go. Carl and I inhale our lunch—sandwiches, cookies and trail mix. When we finish eating, we kick back on the log and breathe in Great Piece Meadows.

I tell Carl about my trip up Brazil's Rio Negro River in 2006. How we went during rainy season, the roughly three-month period between January and March when tropical torrents bloat the Amazon River and its tributaries. At that time of year the river rises as much as 30 feet, inundating the rainforest. My shipmates and I got drenched a few times by sudden downpours, but mostly we enjoyed sunny skies and temperatures that, while by no means cool, were much more bearable than they would have been had we visited during the region's torrid dry season. But the big advantage of being on the Rio Negro during rainy season was the access. Every day we took our 16-foot wooden launches out into the flooded rain forest. Using the outboard at first, and then paddles and machetes and finally our hands, we threaded the small green skiffs through the tops of now mostly submerged trees, flushing psychedelic birds and dog-sized iguanas and the occasional Caiman, a small South American crocodile. The rainforest floor was underwater, far, far below us. We were flitting through its magnificent canopy like birds in boats.

The paddlers who joined Carl on his 2003 trip down the rain-swollen Passaic had a similar experience in Great Piece Meadows. Carl's 20-member group entered Great Piece in a downpour. The Passaic was everywhere. "Once the water tops the bank, the river spills out for a half mile in all directions and loses its velocity," says Carl. "It becomes a very peaceful, gently flowing plain." A water world.

It took a while before anyone in Carl's group ventured out beyond the river's channel and into the woods. Sean Ryan was the first. A landscape architect with the Union County Parks Department, Sean was an experienced kayaker. He had paddled pretty far downriver and had gotten well ahead of the group. Like everyone else, Sean had a map. Consulting it, he realized that he was heading down one side of a very long oxbow. To save time and energy, he decided to shortcut across the floodplain to the other side of the oxbow. Unfortunately, Sean neglected to inform anyone of his plan. "We realized we were missing somebody," recalls Carl. "There were lots of radio calls, and then everything stops. We all meet. There's rain and we can't see more than a few hundred yards ahead or behind us. We're waiting and waiting. We're not sure what to do at this point."

About 20 anxious minutes passed between the time the group noticed Sean was gone and the time he came paddling back towards his relieved companions. "Sean hadn't lost the river," says Carl. "He had found a new one. He literally paddled a half-mile off the river, in three feet of water, through an old growth forest— well, it's about as ancient as you get in New Jersey. We landscape architects call the path that Sean took a desire line. A worn path in a lawn is a desire line. Sean decided to create his own desire line[3]. He paddled into the mystery of the forest."

Sean's off-river adventure emboldened the rest of the group. "We realized then that we didn't have to stay on the river," says Carl. "We could all do a desire line. At that point we just started barreling through the woods. You could see everyone fanning out, just weaving between the trees." For Carl, this rare glimpse into the interior of Great Piece Meadows was akin to discovering that lush cedar bog hidden deep within the Pine Barrens near his Brick Town home. It was mystical and moving.

"If I hadn't been there, I don't think I could have imagined a place like that in New Jersey," he says. "Great Piece is one of the most underrated, underappreciated displays of beauty in the state. I wasn't prepared for just how undisturbed it was. Instead of a lot of invasive species like

---

3 A desire line is a trail, a shortcut really, that has been worn by humans or animals because it is a path of least resistance, the quickest or easiest route between two points.

we'd seen in the suburban parts of the river, and even to some extent in the Great Swamp, what we saw in Great Piece Meadows was magnificent old growth forest with its uncleared understory and its towering canopy. There were silver maples, pin oaks, large sweet gum, box elders, trunks with diameters of 36, maybe even 40 or 50 inches. You're pretty far away from the highways. Everything drifts out of sight. The pounding rain absolutely overtook every bit of background noise. It was so ancient and primal. That's when we met the turkey, this full-grown male gobbler sitting on this big old stump. We thought he'd fly off, but we realized pretty quickly that he really had no place to go. And it was marvelous because this is what turkeys do in the natural world, in *their* world. It was so peaceful a vision: this placid turkey waiting out this massive flood."

The survival of wild life and wild places in New Jersey fills me with joy and amazement and a deep sense of gratitude. Sitting here in the warm spring hush of Great Piece Meadows, I marvel that someone, somewhere, somehow managed, against what must have been overwhelming odds, to wrestle this morsel of nature from the jaws of Garden State developers. It is a genuine miracle.

The miracle worker in the case of Great Piece Meadows is a man named Robert Perkins, the founder of Wildlife Preserves, Inc. And don't get him started on conservation efforts in the Garden State. "You get into the politics of the last 100 years, which gets very turbid and is extremely unflattering to New Jersey," says Perkins. "Think of a bunch of politicians sitting around belching over their last juicy steak. That's a little unfair, but it'll give you a sense. It's a mess."

Robert Perkins is a charming and elusive character. He is oddly cagey about his age, saying only that he was "alive in the 1930s," and "in college in the 1950s," which puts him somewhere just south of 80. We never meet in person. All our conversations take place by phone. Lucky for me, Perkins always answers his cell phone and despite his busy schedule he always seems to have time for a chat.

He was 11 years old when his parents moved from Greenwich, Connecticut to tiny Essex Fells, New Jersey, a 1.4-square mile village just east of the Hatfield Swamp. He lives in Tenafly now, a modest borough in northeastern Jersey whose 07670 zip code, notes Perkins, exhibits "complete bilateral symmetry." Perkins was a born conservationist. "But I never should have gotten involved in New Jersey," he says. "For a tiny, tiny part of the effort that we have put into New Jersey we could have accomplished vastly more in other places."

That is no doubt true. But it is also true that without Perkins and his conservation efforts, the wetlands of the Passaic River's Central Basin would have gone the way of the Newark Meadows. The tree-lined riverbanks that Carl and I have been paddling past today would likely shelter suburban cul-de-sacs rather than bobcats and bobolinks and the last of the blue-spotted salamanders.

Wildlife Preserves, Inc. (a.k.a. Wildlife), Robert Perkins' Newark-based organization, is a self-described "private, nonprofit conservation corporation dedicated to the preservation of natural areas and open space for conservation, education and research." Perkins officially registered Wildlife Preserves in 1952, the year I was born. But his interest in preserving wildlife habitat began much earlier. At age nine, while he was strolling with an aunt through the Garden District of New Orleans, his father's hometown, Perkins shared his sober and prescient concern that the human population boom was threatening to destroy the world. "I realized three things as a kid," he says. "One was the rapidly growing population. Another was the growing power of small groups and individuals with technology. And the third was human nature. You put all three together and it's disaster. At a very early age I was very pessimistic about the future. But I figured I still ought to do the best I could."

Shortly after World War II, while he was a teenager attending boarding school in Vermont, Perkins convinced a handful of wealthy patrons to help him save some wildlife habitat. "They were fat cats," says Perkins, about Wildlife's original benefactors. "People that I knew about, or that some other friends had suggested." Perkins' father was a partner in one of Wall Street's oldest and most prestigious law firms. The Perkins family wasn't rich, but young Robert knew his way around the halls of power. "I was not very fat," he says. "I wasn't fat at all actually. But I got several people interested."

By far the wealthiest patron Perkins courted was Mrs. Marcia Brady Tucker of Park Avenue in New York City. Marcia Tucker was the daughter of Anthony N. Brady, an Albany businessman who in partnership with Thomas Edison helped to found the Consolidated Edison and Union Carbide companies. When he died, Anthony Brady reportedly left one of the largest personal fortunes ever amassed. "Mrs. Tucker inherited a good deal of money," says Perkins. Mrs. Tucker was also a lover of birds, and she had a keen interest in protecting them. Both privately and through her Marcia Tucker Brady Foundation, she became a principal patron of the American Ornithological Union. She sponsored exhibits at the American Museum of Natural History, entertained visiting ornithologists at her homes in Manhattan, Mount Kisco, New York and Florida, and served as a director for the National Audubon Society. In short, Mrs. Marcia Brady Tucker of Park Avenue in New York City was a prime candidate for a fundraising call.

"I remember the first time I met Mrs. Tucker," says Perkins. "I was in my late teens. She had the only freestanding house on Park Avenue. I went there with a friend of hers. We walked in, and there was the butler, who greeted us, and three footmen. They each had these striped dickies and rows of brass buttons down the back of their tails. This was just for a casual visit in the afternoon. I'd been to a number of formal places before, but I was still amazed."

Marcia Tucker died in December 1976 at the age of 93. The Park Avenue mansion that so dazzled the teenage Perkins has since been razed to make way for an apartment building. But through her generosity, Marcia Tucker's spirit lives on in Great Piece Meadows.

Perkins and his investors decided to focus their acquisition efforts on wild areas within 200 miles of New York City. For guidance about which tracts were most worth preserving, they turned to the U.S. Fish & Wildlife Service, the federal agency that has been the steward of America's national wildlife refuges since 1939.

Strange as it seems, there was no one date or event that compelled America to set aside habitat for wildlife, and no single person. The first federal effort to protect a natural area was the Congressional Act of June 30, 1864, which handed Yosemite Valley over to the State of California. Four years later, in 1868, President Ulysses S. Grant preserved the first federally-owned land when he took action to protect the fur seals on Alaska's Pribiloff Islands. The preservation movement picked up considerable steam during Teddy Roosevelt's turn-of-the-century administration. Roosevelt was a devoted hunter and outdoorsmen. His first conservation move was to declare Florida's three-acre Pelican Island part of the public trust in 1903. Today's Pelican Island National Wildlife Refuge is considered the nation's first true "refuge." By the time Roosevelt left office in 1909, he had signed 51 Executive Orders that established wildlife preserves in l7 states and three U.S. territories.

The impetus for Roosevelt and for most federal set asides was a desire to protect the nation's wildlife. During the last half of the 19th Century and into the early 20th Century, America's birds and mammals were being hunted to extinction for their fur, feathers and eggs. Establishing sanctuary areas, where hunting was prohibited, seemed like a good start. But the original isolated pockets of nature weren't large enough or contiguous enough to sustain migratory populations, which can travel thousands of miles. Besides which, preserving land by ad hoc executive fiats was not a sustainable public policy. If America really wanted to protect its wild lands and creatures, the country needed a bigger, better, more coordinated vision.

In 1918 and again in 1929, the federal government made several attempts to expand protections for migratory birds. But the National Wildlife Refuge System that we know today, with its familiar flying goose emblem, didn't really begin to take shape until 1934, which turned out to be a watershed year for conservation.

Congress passed two pieces of landmark legislation that year. One was the Fish and Wildlife Coordination Act, which authorized the federal government to acquire wild lands that would be managed by the U.S. Biological Survey, a forerunner of the Fish & Wildlife Service. The other legislative action was the Migratory Bird Hunting and Conservation Stamp Act (a.k.a. the Duck Stamp Act), which authorized sales of a duck-hunting stamp to pay for the acquisition program. Also in 1934, the newly-elected president, Franklin D. Roosevelt, appointed a blue-ribbon panel to offer advice on how to protect America's waterfowl. Members of the so-called "Duck Committee" were its chair Jay Norwood ("Ding") Darling, Thomas Beck and Aldo Leopold. Never before or since has the cause of conservationism been championed by such eloquent emissaries with such a public platform.

"Ding" Darling was a Pulitzer prize-winning editorial cartoonist who promoted conservation with his syndicated drawings for the *New*

York Herald Tribune. (After his Duck Committee days Darling went on to head the U.S. Bureau of Biological Survey and found the National Wildlife Federation.) Thomas Beck chaired the Connecticut State Board of Fisheries and Game and was an editor for *Collier's Weekly*, the groundbreaking investigative news magazine of the day. Aldo Leopold, the visionary ecologist and U.S. Forestry Service veteran, was a professor of game management at the University of Wisconsin in Madison. Like a modern Greek chorus, Darling, Beck and Leopold warned the public and the president about the growing dangers of habitat destruction and the pressing need for strategies and resources to combat it.

In the years following 1934, the government acted on many Duck Committee recommendations. The Bureau of Biological Survey began to steadily accumulate land around the country for the purpose of creating national wildlife sanctuaries. In 1939, its successor agency, the newly minted U.S. Fish & Wildlife Survey, purchased a labyrinth of salt marsh, coves and bays in Brigantine, New Jersey, a coastal community just 10 miles north of Atlantic City. The former Brigantine National Wildlife Refuge is now part of the larger Edwin B. Forsythe National Wildlife Refuge, 46,000 acres of coastal wetlands and woodlands in New Jersey's Atlantic, Burlington and Ocean counties. Today, the U.S. Fish & Wildlife Service manages 96 million acres of protected habitat in 548 refuges spread across all 50 states. But there was at least one refuge that got away.

Fish & Wildlife officials harbored ambitious plans to create a Passaic Valley National Wildlife Refuge in the 30-mile-long basin of ancient Glacial Lake Passaic. "They had made maps for acquisition, set a boundary and had all the various landowners listed," says Robert Perkins. Then World War II broke out and acquisition efforts ground to a halt. When the war ended, attitudes about conservation and open space began to change. "The GIs were coming home and having families," says Perkins. "Big families." With affordable G.I. Bill loans, those veterans were able to buy homes and start businesses. In New Jersey, the push to develop land began to overtake any interest in protecting it.

In 1956, the U.S. Congress passed the Fish and Wildlife Act. The legislation actually broadened the authority of federal agencies to purchase and develop land for new wildlife sanctuaries. But the bill didn't provide money to underwrite the expansion. Without federal acquisition dollars, the public refuge movement sputtered and stalled. So when Perkins and his investment syndicate approached U.S. Fish & Wildlife officials for recommendations about which properties to purchase, the agency was thrilled, recalls Perkins. "They urged us to get both Troy and Piece meadows."

Even with the benefit of private dollars and the support of U.S. Fish & Wildlife, acquiring the properties proved to be a long and arduous process. Both Great Piece Meadows and Troy Meadows were a checkerboard of tiny narrow lots. "The biggest piece was only about 50 acres and that was unusual," says Perkins. "One plot was 100 feet wide and 9,000 feet long. We had the idea of putting a super bowling alley there."

Wildlife Preserves, Inc. spent the 1950s and 1960s tracking down and negotiating with more than 100 different landowners. "Some of them had owned farms in the 1800s or 1700s and the farm would buy a little tract in the meadows as a source of timber and bedding hay, maybe only two or three acres," says Perkins. "Except along the roads on the edge of the meadows, none of the owners lived there. Some owners had just disappeared."

Despite the difficulties, Wildlife managed to round up an impressive share of the basin that once contained Glacial Lake Passaic, including 2,500 acres in Troy Meadows and 500 acres in Great Piece Meadows. (The organization currently owns about 6,000 acres throughout New Jersey.) Wildlife was only able to grab fragments in Great Piece Meadows, a small tract here, a small tract there. But in the long run the fractured nature of the group's holdings turned out to be a strategic advantage. "The town of Fairfield wanted Great Piece Meadows developed," says Perkins. But every time town officials tried to build in Great Peace Meadows, some little wedge of Wildlife land got in the way of their plan. "By being scattered around," says Perkins, "our property prevented any major development." In Great Piece Meadows, that is.

Continuously thwarted in its efforts to build in Great Piece Meadows, the Township of Fairfield drained and developed Little Piece Meadows instead. That decision haunts local property owners, taxpayers and the Federal Emergency Management Agency every time a big storm swells the Passaic. The Willowbrook Mall in nearby Wayne, New Jersey was built on top of Little Piece Meadows. The 200-store retail temple is one of New Jersey's oldest and largest shopping centers. In August 2011, torrential rains from Hurricane Irene left every store in the mall—and most of Fairfield Township—under four feet of water.

Wildlife Preserves, Inc. hasn't been able to shield Great Piece Meadows and Troy Meadows from all the forces that threaten their ecology. The road salt that bordering towns heap onto local streets and highways in wintertime has begun to alter the freshwater character of the wetlands. In many places, the invasive reed *Phragmites*, which can tolerate salt, has replaced the once large stands of cattails (*Typha*), which cannot. Ditches dug to drain standing water and control mosquitoes are lowering the natural water table in the meadows, which may explain the decline in the populations of black ducks that feed and nest there.

As Wildlife's Land Manager, Leonardo ("Len") Fariello worries about these and other assaults on the meadows' fragile ecosystem. "The biggest threat is human beings and human activity," he says. "When I was growing up here more than half the land was vacant. It was just awesome. It all changed so fast. Who knew it was going to be covered over with all this suburbia?"

I drove out to Len Fariello's house in Whippany on a drizzly July morning. He had agreed to take me on a tour of Troy Meadows and Great Piece Meadows. "We're going to get wet," he said, as he opened his front door.

Len is short and dark and handsome. His thick dark hair was slicked back. He wore cut-off blue jeans, a black t-shirt and a pair of beat up running shoes. He had a pinky ring on his left hand and a diamond-studded gold band on the ring finger of his right.

He led me back into his kitchen, which was crowded with rocks, fossils, Native American artifacts, antique telephones and dozens of vintage glass milk bottles displayed on a high wraparound shelf. Old pint milk bottles are a passion of Len's. He fills them with white Styrofoam beads to make the logos from the long-gone local dairy farms pop. "I'm a collector," he acknowledged: "of everything." Including the three dogs yipping and scratching at the outside of the kitchen door— and the five orphaned raccoons, three baby skunks, 'possums and grey squirrels that share a fenced enclosure in the backyard.

Len unfolded several large maps of Great Piece Meadows and spread them out carefully on the kitchen table. I could clearly see the property lanes that Robert Perkins had described. One map, a plat map of Great Piece Meadows prepared by the Army Corps of Engineers, showed the slim slats of land tucked together tongue-and-groove like floor planks. "In Fairfield Township, one half of Great Piece Meadows is designated a wildlife management area, the other is a wildlife sanctuary," said Len, pointing out the two sectors on the map. Hunting is allowed in the wildlife management area, which is mostly owned by Fairfield. That's where members of the Fairfield Conservation and Sportsman's Association hunt. But hunting was and is illegal in the sanctuary areas owned by the New Jersey Department of Environmental Protection and managed by Wildlife Preserves, Inc. Part of Len's job is patrolling the boundaries where the two zones meet.

We set off in Len's green Chevy truck, the maps and a portable GPS wedged between the dashboard and the windshield. We drove north for about ten minutes then turned east onto Troy Meadow Road, a rutted, grass-covered trail that runs straight through the middle of Troy Meadows. The Passaic was about seven miles to the east. The night before, severe thunderstorms dumped nearly two inches of rain on this part of northern New Jersey. There was standing water everywhere. When we reached the spot where West Brook joins Troy Brook, just a few yards to the north of Troy Meadow Road, the passage was flooded. We bumped and rolled and sloshed through swirling creek water, which in places covered the wheel wells. We were moving through a forest of young trees. The foot-high grasses that carpeted the road and the forest floor were neon green.

Len stopped the truck beside a red metal gate. The heavy gateposts were bent in opposite directions. The gate had come unhitched from the right post. That end rested a kilter on the ground. We stepped over it easily and walked a short distance through a grassy tunnel of trees that opened into a small clearing in the woods. "It was right here," said Len, as he stood on a low mound at the edge of the clearing closest to the truck.

The mound was the site of the burned down shack that Len was squatting in when he got busted by wildlife rangers back in September 1973. The rangers knew Len. He was the local nature boy from Whippany, a regular Huck Finn. He and his childhood pals used to run a raft up and down the Whippany River. (The Whippany tracks the eastern edge of Troy Meadows on its way to meet the Passaic.) September is the start of New

Jersey's duck hunting season. The Wildlife rangers figured they could use an extra hand, so they offered Len a job keeping poachers off the wildlife preserves in Great Piece and Troy Meadows.

"I went by the name of Running Deer back then," said Len, who in his younger days was preternaturally fleet. He was New Jersey's high school Cross-Country Champ in 1965. "I was a bare-footer, except in the dead of winter. The bottoms of my feet were calloused like leather." Running Deer was a lot faster and a lot quieter than his typical prey, the hunters in their clumsy rubber boots or hip waders. He bagged some 60 poachers in the Troy Meadows sanctuary that first season.

Len stayed on with Wildlife for the next three years. He fixed up the cabin, scrounging a wood stove, and installing a screen door and windows and a skylight where some vandal had blown a shotgun hole through the roof. "I slept in a loft under the moon and stars," he told me. "It was pretty primitive, but that's how my life was back then. I was just playing." Len was serious about his work though. During his time as a ranger, Len vigorously enforced the hunting ban on Wildlife property, waging war against poachers and becoming something of a legend in the process.

He was uncanny in his ability to track poachers and relentless in his pursuit. It was common for him to go crashing through stands of cattails on the heels of fleeing deer and duck hunters armed with shotguns or long-bladed hunting knives and high-powered bows. Not surprisingly, Len endured his share of retaliation over the years. He's been punched and showered with birdshot. He was knocked down and run over by one group of angry violators when he refused to back away from the grill of their jeep.

"I had a hunter draw his bow on me once," Len recalled. The bow incident happened on a chill, damp autumn afternoon when Len and a partner were out patrolling the edge of the swamp. Len noticed some fresh boot tracks in the mud. "Most bow hunters hunt from trees," he said. "I probably wouldn't have seen this guy if he had just stayed still and hugged the trunk, but he panicked, scrambled down the tree and started running."

Len took off after him. After a half mile run through deep water, Len caught up with the hunter at the edge of the woods behind some houses. Then all hell broke loose.

"He notches an arrow and draws back on me with a four-blade razor point aimed at my chest," recalled Len. "He's in full camo, including face paint. He's yelling, 'Stay back! Stay back!' Dogs start barking. Girls start screaming. I'm yelling call the police! Call the police!"

In all the commotion, the bow hunter managed to duck into one of the nearby houses. Local police eventually negotiated a surrender. "Turns out it was a notorious trapper and poacher I'd tangled with before," said Len. "I never recognized him that day with all his camo and hunting regalia."

All in all, Wildlife Preserves, Inc. successfully prosecuted about a third of Len's collars. "I sort of cleaned up the area," he said. And then, in 1976, he left New Jersey.

Len wound up in Arkansas' Ozark Mountains. He bought some land and took up farming, raising beef cattle and cultivating crops and medicinal roots and herbs. The locals knew him as "Len Sunchild" or "Doc." Bob Perkins got in touch every September, to see if Running Deer would come back and help out during hunting season. Most years, Len would go. "I never broke my connection with Wildlife Preserves," he said. It was a good thing too, because after a decade in the Ozarks, with no TV, radio, telephone, electricity or company, even Len was ready to "get back to society."

In 1986, he returned to New Jersey and to fulltime work with Wildlife. He married a Whippany girl, had two kids. Running Deer domesticated. Len and his wife are divorced now. But he still lives in Whippany with his daughter and son, and he still works for Wildlife. Len has been the organization's Land Manager for more than a decade.

Each fall, when he would arrive in New Jersey from Arkansas to work the hunting season for Wildlife, he'd be struck anew by the changes. "All the places I played in as a kid were being paved over and developed," he said. The dramatic changes moved Len to write books about his youthful adventures (*The Nature of Changing Times*) and about his once rural hometown. *A Place Called Whippany: The History and Contemporary Times of Hanover Township, New Jersey*, was first published in 1998. Len dedicated the book to his daughter Lydia and son Luca and "To the children of the future and the spirit of the past."

It seems that every New Jersey nature lover has some wrenching tale about a desecration of the natural world. A woodland bulldozed, a river dammed, a pasture paved. For the storyteller, the assault is often so brutal and so careless that it has the feel of terrorism. Len Fariello's tale involved an old beech tree that grew behind his childhood home.

Len grew up on the edge of Whippany Farm, a 138-acre estate originally owned by the Frelinghuysens, a prominent Morris County family. His backyard was a lush undisturbed forest with towering stands of sycamore, oak, tulip and hickory trees and one magnificent grove of virgin beeches that occupied an oxbow of the Whippany River. One of the oldest and most graceful beeches stood right behind Len's house.

"It had an enormous five-foot diameter trunk," said Len. "Five kids with joined hands could barely encircle it. There were 'possum and 'coon dens in it, and song bird nests and a wood duck nest in the top, and my friend and I built a tree house in it, way up high where only we could climb."

In 1972, when Len was a teenager, the state of New Jersey began work on an extension of Route 287. For the northernmost stretch of highway, the state appropriated portions of Whippany Farm and of the Fariello property. "When they built Route 287, they just cleared everything," said Len. The backyard beech was among the thousands of trees felled in the process. "When they destroyed that forest," he said, "I snapped."

With the trees gone and his family displaced, Len eventually hopped on his Harley and spent the summer of '73 traveling the country to California and back. When the Harley broke down in Chicago on the way home, Len and his disabled bike hitched a ride back to New Jersey. With no money and no prospects, he took shelter in the loft of the old abandoned hunter's shack in Troy Meadows, which is where the Wildlife rangers found him that September.

Len and I stepped gingerly around the old cabin site, careful to avoid the bricks and twisted cable and chunks of old rusted bed frame that poke up through the grass. We were looking for the footprint of the old fireplace. We found a female turkey instead. She was sitting on her nest, little more than a shallow depression in the tall grass. We stopped as soon as we spotted her, but she startled and flushed anyway, leaving her eight eggs behind. They were pale beige with brown speckles and surprisingly small, no bigger than large chicken eggs. When I chided the mother turkey for abandoning them, Len sided with the bird. "She probably made the right move," he said. "She'll be back."

Before we left the old cabin site I snapped a picture of Len standing on the mound near what used to be his old front door. Knee-deep in the wet grass, with the forest at his back, Len stood up straight with his hands at his sides and looked directly into the camera. His strange half smile betrayed a trace of awkwardness at having his picture taken, and a hint of mischief, nostalgia, defiance and something else. It's hard to say exactly. It was just Lenny, a John Muir/Joe Pesci mashup. The Jersey tough guy who can get all misty-eyed about bog turtles.

"At first, I ran away from the development," said Len, as we walked back to the truck. But eventually, "I decided to take a stand and fight it." When he returned from the Ozarks in 1986, Len formed a nonprofit group called Citizens for Controlled Development. He began publishing a newsletter that profiled natural areas threatened by development and offered practical tips on how to contain and manage growth. It wasn't enough. "I was always arguing my case to the people who were in control," he said. "I realized I had to be on the other side of the podium." He decided to run for public office.

It took a few tries but in 1999, campaigning on a "Save It, Don't Pave It" platform, candidate Len Fariello won his first political race— for councilman in Hanover Township. He served on the council for 10 years before retiring from politics in 2010. During his tenure, Len helped Hanover acquire more than 300 acres of open space, and sponsored several ordinances to protect natural areas and native trees.

In his dual careers as Wildlife Land Manager and local pol, Len came to appreciate the art of compromise. He's been trying (unsuccessfully) to convince his boss, Robert Perkins, to develop parcels of Wildlife land for recreational use as a way to pay the bills. But Len is a pragmatist by necessity. At heart, he still fights for every stream, every tree, every turtle, bird and bug, and he mourns the loss of every spec of nature. Which is what makes the land and wildlife of Troy Meadows and Great Piece Meadows all the more precious to him.

On the drive from Troy Meadows to Great Piece that afternoon, Len spotted a dead mink on the side of Route 80. He pulled over to retrieve the body, stiff but intact. "We'll find a better resting place for him," said Len, tossing the mink into the bed of his truck. "When I take Cub Scouts and Boy Scouts into Great Piece Meadows or Troy Meadows, I tell them that this is the land for animals, not people. We're entering their domain.

People just don't care about wildlife. The county, the state, they just want to take our land. They have no plan in place to protect and manage it. The Army Corps wants it for flood storage. The towns want it for recreation. They seem to be more concerned with human use, with grass and picnic benches. Their attitude is that it has to be improved."

In 2006, Wildlife Preserves, Inc. sold 400 acres of its Great Piece property to the Army Corps of Engineers. The Corps, acting on behalf of the State of New Jersey, has been buying up Passaic River floodplain property as part of the so-called Passaic River Natural Flood Storage Area Project, a flood control program that was ratified by state legislators in 1999. Under the terms of the Great Piece sale, Wildlife continues to manage the property in partnership with the New Jersey Natural Lands Trust, a division of the state's Department of Environmental Protection. But the Great Piece Meadows land is now subject to a "flood control easement," explains Len, which means that the Army Corps can do pretty much whatever it wants in the name of flood control. "If you read the easement, it'll scare the hell out of you. They're buying the land for flood storage, but there's all kinds of language where they can do a lot to manage floodwaters, [including] build dams and dikes, change the hydrology, control the water."

Great Piece Meadows, as it exists today, may be one big flood away from an Army Corps makeover. Erecting dams or digging ditches may help to mitigate flooding in the Passaic's River Valley. But such interventions are certain to disrupt the delicate balance of these Central Basin wetlands.

It's almost 2pm when Carl and I get up from our lunch log and scramble back down the bank to the Pungos. Across the river, a couple is stepping out of a canoe and unloading their own picnic. They are the first people we've seen since we paddled into Great Piece Meadows some four hours ago.

We wave and holler greetings back and forth across the channel before setting off downriver again. The sun is still high in the sky. The river's surface is a blend of all the colors that surround it, as if a painter had mixed the dollops of paint on his palette into a sparkling greenish brown.

Before too long we see another boat. This one is a 16-foot skiff with an outboard that it is ferrying four young men upriver. Each of them holds a can of Budweiser. The one in the bow has a chainsaw at his feet. They nod grimly as they approach us, and politely throttle back their engine so its wake doesn't swamp the kayaks. This must be one of the self-appointed chainsaw gangs Carl told me about, groups of (mostly) guys who cruise the river cutting through strainers to keep the Passaic passable for motorized boats. Where were they yesterday when we needed them? I consider alerting the strainer patrol to that wide, tricky logjam upstream. But then I reconsider. If that collection of timber hadn't been blocking the river I never would have witnessed the Roy Rogers. Out of respect for Carl's best move, mum's the word to the chainsaw crew.

Not long after they disappear around a bend, I catch a flicker of movement along the left riverbank. A long black shadow darts behind a maple that is clenching the bank. The tree's root ball is completely bare. A single poison ivy runner twirls up its trunk. I alert Carl. We paddle slowly and quietly towards the shore. When we get within 10 feet, we stop paddling and hover, keeping our eyes on the tree. Suddenly, a small black face peers out from behind the trunk and then just as quickly withdraws.

"That's a mink!" exclaims Carl. He sounds as shocked as I was when Len found the dead mink on Route 80. Before that, I'd assumed the only minks in New Jersey were on hangers in the Short Hills mall. We let the boats slide closer until their bows just nuzzle the bank. I figure the mink is long gone, but after a few seconds it pokes its head out from behind the right side of the tree, then from the left, then the right again. Then the antic mink scrambles around the tree, right in front of us, and disappears into the root ball.

"Is that his nest?" I ask Carl in a whisper.

"Den," says Carl. Minks make their homes in abandoned burrows along riverbanks, or in hollowed out logs, or old beaver lodges, or in a ball of exposed tree roots just like this one. They are opportunistic homemakers.

I consider rummaging in the dry sack for my camera, but before I can make a move a wild thrashing and billows of dirt erupt from the root ball, and out flops a fish, a large fish, nearly as big as the mink itself, which pounces out after it. A smack down ensues right there on the riverbank, not five feet in front of us. Mink versus Fish. They flip and spin and twist and tumble back and forth along the bank, a ball of sleek black mink and grey green fish, faster and faster until they dissolve into a cartoon blur. Suddenly, the mink loses its grip. The fish breaks free. He makes for the river, but before he can slip back into the Passaic, the mink is on him again, this time for good. With his jaws locked around his vanquished opponent, the mink trots upstream, no doubt to some backup lair where he can eat his prize in peace. The victorious mink casts a final backward glance at Carl and me, then ducks off the path and into the underbrush.

This episode is why I love kayaks. Maybe Carl and I would have gotten front row seats for the mink spectacle no matter what kind of boats we were traveling in. But I think the Pungos had a lot to do with our luck. Kayaks are just so quiet on the water, even big ones like these. There's no engine whine, no squeaky oarlocks, no thudding of paddles against gunwales, no loud unnatural sounds to signal your approach. In a kayak, you can surprise wildlife. And even after the creatures you're sneaking up on finally spot you, they don't always dash off. The kayak's aspect in the water makes it less threatening than motorized boats or even canoes. Kayaks sit low. They don't loom. They blend in. When you're in a kayak you become a water creature, a mermaid with a kayak for a tail.

The mink will be our best, though not our last encounter with Passaic River wildlife. Carl and I have spent some 15 hours on the river at this point. We've traveled about 30 river miles. Camp Lane is five miles or so downstream, which means we have about three more hours of paddling left before we exit Great Piece Meadows. Once we reach the take-out, we'll leave this last wild stretch of the Passaic behind. Beyond Camp Lane the river enters the Lower Valley where its banks will be more inhabited and its waters less clean. "The Passaic now clearly becomes an urban river," writes Gertler about the reach below Two Bridges, "It will appeal to few…" I take umbrage at that description of my stretch of Passaic, but I can't really disagree.

There are still wonders of nature to come. The thunder of Great Falls and Little Falls awaits us downstream. But nothing ahead will compare to the peace and, for the Passaic at least, the vast emptiness of this secluded wetland in the northern part of the Central Basin. Having paddled Great Piece Meadows, I understand Robert Perkins' desire to preserve these wetlands and Len Fariello's passion to protect them. Great Piece Meadows has "amazing atmosphere," says Carl. It is the Passaic's final unfettered fling before being dammed, diverted, bridged and bulkheaded beyond recognition. Before it is tamed and broken.

Just before we reach Camp Lane, Carl and I run into a young family paddling upriver in a big red canoe. They are the last people we'll see on the river today. Dad is in the stern. Mom is up front. Their two dark-haired daughters—I'm guessing three and five years old—are sharing the middle seat, crunching on snacks from a white plastic container that the older sister is clutching in her lap. The girls munch in silence, but their giant brown eyes follow Carl and me as we maneuver our orange Pungos alongside the red canoe. Their father is a wildlife photographer. He's never canoed the Passaic before, he tells us. But while he was out on assignment yesterday, he happened upon a pretty boat launch just upstream. This morning he decided to take the family out for a Sunday paddle. "It must be beautiful in the fall," he says, motioning to the tree-lined banks.

The photographer isn't worried about exposing his family to toxins, because there aren't any—or many—to speak of on the upper Passaic. The river is especially clean here in Great Piece Meadows. I imagine that the impressions that the photographer's two young daughters are forging right now of the Passaic will be quite different from my own. They won't remember a darkly menacing slough. Instead, the Passaic will conjure this placid paddle along tree-lined shores with their parents on a sunny Sunday afternoon. Carl and I sow one more memory seed. We wow the family with our mink story before saying our goodbyes and continuing on our separate ways—Carl and I downstream to Camp Lane, the photographer and his family upstream into the quiet mystery of Great Piece Meadows.

When we reach the take-out, we make quick work of pulling the Pungos out of the river and loading them onto his car. When all the gear is sorted and stowed, I thank Carl for all his logistical legwork, and for keeping us safe, and for being such an intrepid and cheerful companion. Although eager to get to the Communion party at my brother Joe's house, I'm a little reluctant to leave. It's been an amazing two days. It feels like Carl and I have been traveling together much longer than that, which I take as a good sign, a measure of our paddling compatibility. We finally hug goodbye and climb into our cars. We'll reconvene five months from now, in October when we paddle the last 30 miles of river.

I trail the Camry with its two orange Pungos secured aloft. We crawl down Camp Lane to Route 46 East to the Garden State Parkway South.

Once we hit the Parkway, I zoom past Carl, who is hanging in the slow lane. It's almost 6pm. I call Joe from my cell. He picks up on the second ring. I can hear the chatter of guests in the background. "Keep the party going," I tell him. "I'm on my way!"

7 | LITTLE FALLS

SIX MONTHS LATER, IN MID-OCTOBER, Carl and I are ready to get back on the river. We get an early start on a crisp Saturday morning. So early in fact that we have to spend 10 minutes scraping a hard frost off the car windshields before we can set out for the day's paddle. The whole calculus of the cars—the obligatory drive to the take-out, then back to the put-in—is especially maddening on this crisp autumn day. As I tail Carl from the Holiday Inn in Totowa, New Jersey where we are staying, southeast to Memorial Park in nearby Fair Lawn, where we will leave my car for the day, I keep getting glimpses of the Passaic through the trees. The river's surface is like glass. I just want to get out on the water before the wind picks up and puckers that seductive early morning stillness.

The Passaic we travel down today will be a new and different river. No more winding lolls through lush empty wetlands. The river has work to do now, and so do we. After flowing north astride the Watchungs for more than 40 miles, all the way from Millington, the Passaic will finally break free of the mountains. In the next six miles, the river will explode through the two remaining Watchung ridges, then pivot sharply south and break for the sea. The placid river of Great Piece Meadows will turn wider, faster and fiercer, coursing over back-to-back cascades as it blasts through the basalt. The first, smaller drop at Little Falls is just two miles downstream from our launch point. The second comes four miles later, in the middle of Paterson, where the Passaic leaps from a towering cliff at Great Falls. The day promises a little whitewater paddling, two strenuous portages and a lesson in the power of persistence. In this stretch of river, the irresistible force of the Passaic will meet and finally conquer the immovable Watchungs.

Cathy Elliott-Shaw is waiting for us at Memorial Park. Cathy is an old friend of Carl's who is coming along on this leg of the trip. She is small and trim with short brown hair and a big smile. She seems like a gamer, the easygoing, pitch-in sort who will help carry your boat or pick up the tab or shove you off so you don't get your feet wet. Cathy works for New Jersey's Green Acres program. She heads up what's known as the Urban Acquisition and Statewide Park Development Team. This team helps counties, cities and nonprofit groups purchase and preserve land for conservation and recreation. On occasion, Cathy's crew acquires land for state parks and wildlife areas. All these Green Acres acres are potential restoration candidates for Carl. They will paddle together today in the royal blue Perception, a double kayak that is lashed to the roof of the Camry alongside my trusty orange Pungo.

We make a quick pit stop at the Fair Lawn Dunkin' Donuts for coffee and pastries. By 9:30am we're crammed into Carl's Camry heading west towards Camp Lane, our favorite boat launch. The temperature is climbing with the sun and everything is going swimmingly. Twin falls awaits us! And then we hit a monster traffic jam on Route 46 just outside Totowa.

Cathy, who is riding shotgun, starts thumbing through the two maps on the dashboard in search of an alternate route. As the snarl of cars and trucks in front of us slows to a crawl, Carl tries to help, jabbing at the map Cathy is holding, tossing out options and then shooting holes in his own suggestions in a geographic dialectic worthy of Socrates— or Robin Williams. There are some people who can't abide sitting still in an automobile, and Carl is one of them. When the blinker on the NJ Transit bus ahead of us starts to flash, Carl decides to take a flier. "That bus driver knows something," he mutters. With a quick glance over his shoulder, Carl lurches into the right lane after the bus.

I'm not exactly sure how we made it to Camp Lane that morning. I'm not sure Carl knows either. Our detour involves a few twists and turns and a newly-paved, two-lane asphalt road that snakes west through a wooded area north of the highway. Somewhere along the way we lose the bus. But 10 minutes after abruptly exiting Route 46, we are pulling into the familiar little gravel parking area at the end of Camp Lane. It's 10:30am.

We speed sort the gear. Extra paddle, bilge pump, safety rope, life jackets, first aid supplies, clothes, cameras, car keys, wallets, cell phones, snacks, more rope, water bottles, walking shoes, maps, wheelies for portaging. Check. Check. Check. Check. Check. We stuff it all into dry sacks and tamp them into the bow and stern of the boats. It's almost 11am by the time we finally shove off. That's a late start given that we have to portage around Little Falls and Great Falls today. But we're in luck. After two days of heavy rain earlier in the week, unusual for October, the river is running fast. The current will help us make up the lost time.

I paddle slowly backwards, away from the launch, and wait for Carl and Cathy. Autumn has tinted the trees along the banks here at Camp Lane. Reflected dabs of red and gold brighten the river's drab army green surface. From the river, Old John's house looks pretty much the same as it did when Carl and I were here last. Plywood still covers the two first floor windows. The menagerie of vehicles still clutters the yard. The place still has that soft-focus kind of timelessness, like a cherished memory. Len Fariello would love it here. I have to remember to tell him about it.

From Two Bridges it's an easy two-mile paddle to Little Falls. The late morning sun is warm and the river is wide enough here that we can paddle in sunlight the whole time. Three red-tailed hawks circle overhead. Now and then a puff of air riffles the surface, but there's no wind to speak of, and no strainers either, nothing to impede our progress downstream. The autumn air is bracing, the water is smooth and the current is going our way.

This twin falls section of river will take us clockwise in a broad semi circle. We'll paddle north and east at first, past Two Bridges, where the Pompton joins the Passaic. Further along we'll negotiate the gaps in the middle and easternmost Watchung Mountains— at Little Falls and Great Falls, respectively. A few miles downriver from Great Falls, a snout of river marks the northernmost point in the Passaic's 86-mile run. At that high point, between Paterson and Haledon, the river will curve around to the south and we'll follow it into the Lower Valley and our take-out at Fair Lawn. Total distance of today's paddle: about 12 river miles.

Within minutes of launching at Camp Lane we sweep past Two Bridges and the mouth of the Pompton. They come up so quickly I almost miss these important landmarks. Carl has to alert me to turn around and have a look.

The Pompton River is the Passaic's largest tributary. It rushes downhill from the highlands of northwestern New Jersey, gathering water and power from its own mountain stream tributaries along the way. All that volume bursts into the Passaic River here at Two Bridges. The Pompton surge is one of the principal culprits behind flooding both upstream in the Passaic's Central Basin and downstream in the Lower Valley. The Passaic swells with the incoming flow. It is, and with one exception will remain, a larger river now, nearly four times as wide as it was at Camp Lane. I bring my boat fully about to witness this wedding of the waters.

The brief turbid union of the Pompton and Passaic forms a perfect upper case Y.

The two bridges that cross the two rivers here are low, simple structures, which connect to each other at a slightly obtuse angle. Neither bridge opens, so their infrastructures are spare, unburdened by hulking machinery or soaring steel trusses. They look and feel old timey. The clean line of these bridges marks a navigational halfway point: 42 miles from the river's source, 48 miles from its mouth.

Two Bridges also draws an economic and demographic line across the Passaic, separating the wealthier suburbs upriver from the more working class communities downstream. The change isn't dramatic. As we continue downstream, we slip beneath the Route 80 and Route 46 bridges, curve around the Willowbrook Mall and pass under the Route 23 Bridge. The river is still fairly scenic. Oak and maple trees command the banks, their drooping bows create caves of shade along the shore. The leaves are just beginning to turn. There are few signs of industry. Only the occasional pair of white plastic lawn chairs waiting expectantly in small waterfront clearings heralds the houses that will soon begin to fill the banks.

We are paddling the border between the Borough of Totowa on the river's north shore (to our left), and the Township of Little Falls to the south (our right). Both are small leafy communities with populations of 11,000 and median household incomes in the $70,000 range[1]. The houses along the river here are modest, single-family structures whose owners have claimed their riverfront with a mix of homemade wooden docks and decks and bulkheads of concrete, railroad ties, stone or brick. Along one wooded stretch in Little Falls, just downstream from Two Bridges, someone with a flare for masonry has built a handsome low slate wall at the water's edge. There's a small slate patio on top where a white plastic chair

1 Source: U.S. Census Bureau's 2005-2009 American Community Survey.

and bistro table sit in the shadow of a giant oak tree. At the back of the patio a half dozen wide slate stairs extend up from the river in a shallow arc. We can't see a house from the water; it must be somewhere back up in the woods away from the river's edge. But at the top of the bank, where the stairs disappear into the trees, there is an eight-foot-tall tepee made of branches. Carl spots a second smaller patio with a white chaise lounge embedded in a shady thicket just downstream from the tepee. Whoever built this rustic retreat clearly loves the Passaic.

The riverside above Little Falls was a popular vacation spot at the turn of the 19th Century. City dwellers desperate to escape the summer heat and humidity could hop one of several train or trolley lines and spend the day along the shady Passaic and Pompton Rivers. There were parks for picnicking, beaches for bathing and canoes and rowboats for rent. At one time, when the Morris Canal[2] still provided a water link between the Passaic and the Pompton, it was possible to paddle from Singac, New Jersey—now West Caldwell— all the way to Pompton Lake and back. The distance was about 16 miles overland and much farther by boat, which made the trip a daylong adventure.

The *New York Times* published a detailed, first-person account of this Singac-to-Pompton odyssey in August 1916. "The day was Sunday, the hour was 9 in the morning, the place was Singac, a small hamlet on the

---

2 Considered an engineering marvel, the Morris Canal was created to transport coal, by boat, from the mines of Pennsylvania's Lehigh Valley to industrial centers in northern New Jersey. Construction began in 1825 and was completed in 1836. Chief Engineer Ephraim Beach used mules, aqueducts, 23 conventional lift locks and 23 water-powered inclined planes to move boats 102 miles from the east bank of the Delaware River at Phillipsburg, New Jersey, to Jersey City on the banks of the Hudson River. The Canal passed through many New Jersey towns and cities, including Port Warren, Saxton Falls, Netcong, Boonton, Paterson and Newark. The elevation change from start to finish was a record 1,674 feet, making the Morris Canal the largest "hill climber" in the world. The Canal was decommissioned and drained in 1924.

Passaic River southwest of Paterson," began the writer and paddler Albert Handy. "...We had risen with the milkman and the early worm that we might take an 8 o'clock [Erie Railroad] train from Jersey City."

When Handy and his unnamed female companion reached Singac they rented a canoe at a boathouse that was directly across the river from the train stop. They didn't return it until well after dark. The couple might have made it back earlier if they hadn't lingered so long at their "well-chosen" lunch spot somewhere north of Wayne: "North and east and west lay the purple mountains and between lay the woods and at our feet the stream, and there were no man-made noises and no signs of man-made handiwork except such as we were guilty of having brought with us," wrote Handy.

The riverside communities around Little Falls are more built up today than they were in that summer of 1910. But this part of the Passaic River Valley still retains some of the same wildness and lazy river charm that inspired Albert Handy. Like Camp Lane, the hidden neighborhoods along the Passaic and lower Pompton in places like Fairfield and Little Falls and Wayne Township have a kind of timeless quality. Poking around here it is easy to imagine stumbling upon a young paddler and his girlfriend lazing in the grass on a sultry summer afternoon.

As a way to draw even more daytrippers like Albert Handy, the local train and trolley companies began building amusement parks along the Passaic and other New Jersey rivers and lakes. Idlewild Park opened in 1895 along the Passaic River just below Little Falls. It was one of the premier amusement parks of the day. Idlewild was easy to get to—a convenient stop on the Paterson-Singac trolley line—and there were lots of things to do once visitors arrived. The park offered canoe and paddleboat rentals, swing sets, shady picnic grounds, a baseball diamond, a merry-go-round and a large pavilion for dancing.

Like most of north Jersey's amusement parks, and the railroads and trolley lines that serviced them, Idlewild was eventually done in by the rise of the air conditioner and the automobile. Cool air on demand kept the daytrippers at home. The family car made it possible for the more restless souls to vacation farther afield, and on their own time schedules too. Palisades Park hung on the longest. The home of the world's largest saltwater pool, perched high above the Hudson River in Cliffside Park, New Jersey, survived until 1971. By then, Idlewild had been gone for almost 30 years. Kearfott Guidance and Navigation occupies the old Idlewild site today.

We are getting close to Little Falls now, the day's first cascade. Time to pay attention. We do *not* want to miss the take-out. As it happened, a month after our kayak trip a young man drowned not far from the place where we'll pull the boats out of the river today. His name was Joseph Rivera. He was 27 years old. He lived in Stanhope, a town in northwestern Jersey not far from the Delaware Water Gap. On the night of November 13, Rivera and three friends were having a few beers at a riverfront home in Little Falls. It was an unseasonably warm night, and the four friends decided to go for a canoe ride. They borrowed a neighbor's 17-foot aluminum Grumman and set out.

There had been a lot of rain in November. The river was running higher and much faster than it is today. None of the men was wearing a life jacket. As soon as they got out onto the river the four friends knew they had made a mistake. The water was fast and rough. They tried to maneuver the canoe back to shore, but the swirling current gobbled up a paddle. They lost control. The boat capsized.

Rivera's three friends made it safely to shore. So did the canoe, which washed up several miles downstream. When rescue teams failed to find

Joseph Rivera, they assumed he'd been swept over the falls. They were half right; Rivera was actually swept over two falls. His body was found 12 days later, near the hydroelectric plant below Great Falls, nearly five miles downstream in Paterson. I shudder when I imagine Joseph Rivera being hurtled down an icy river in the middle of the night with the roar of the falls a crescendo of doom. He must have been terrified. Here was the dark side of the Passaic. "It is an angry river," the canoe's owner told the local press.

Carl and Cathy paddle over to me, and as we float along together Carl profiles the take-out for us. We are looking, he explains, for a large grey rock on the right-hand bank by a tennis court next to a condo complex just past an asphalt street that comes down to the water's edge. I'm still digesting these details when we see Grey Rock Avenue dead ahead. The aptly named road is landmarked by two grey boulders that rest low in the water. The surfaces are smooth and almost flat, like docks, and spotted with circles of sage green lichen. They are handsome, chiseled rocks that any landscaper would covet. I back paddle to hold my position in the current while Carl parallel parks the double at the upstream rock. The stout tree branch above it is within easy reach. Cathy grabs hold of the branch and hauls herself up and out of the boat. Carl and I follow her lead.

The bank behind the rocks is flat and dry and grass-covered—a very good take-out spot. The best we'll have all day. We haul the boats out of the river and carry them a short distance down a flat dirt path to a strip of mowed lawn that forms an apron around two empty tennis courts. It's time to unpack the wheelies, the all-important accessories that we'll use to roll the boats through the streets of Little Falls and then Paterson (and tomorrow around the Dundee Dam in Garfield too.)

Carl has brought along two different kinds of wheelie. I get the homemade version that was designed and built by his Aussie friend Gary; two old, hard rubber lawnmower wheels, a plastic axle with a V-shaped saddle for the kayak's keel, and two lengths of rope. Carl has the Roleez model, which he purchased special for the trip: black cloth straps, soft rubber tires and a collapsible aluminum frame. The Roleez advantage, apparent right away, is that unlike my DIY contraption, Carl's set up supports the entire weight of the boat. Which is helpful when you consider that we'll be hauling these 14-foot-long, 70-pound kayaks down busy city streets, through parking lots and construction sites, over curbs and down cobbled paths for, oh, probably a mile or more when you add up the three portages we'll make on this weekend river trip.

With wheels down, we head off, single file with Carl in the lead. Cathy spells me as we roll the boats along the narrow asphalt footpath that winds from the tennis court over a footbridge, through the grassy common area along the river and into the asphalt parking lot behind the old Beattie's Carpet Mill, which is now an exclusive condo complex called simply The Mill at Little Falls.

Ireland-born merchant Robert Beattie built this brownstone carpet mill on Main Street in Little Falls in 1844. The buildings are quite beautiful. They stand three stories high and stretch along the right (or south) bank of the river for a thousand feet. Robert Beattie died in 1882. But his carpet mill operated for another hundred years, until 1982 when the family-run business defaulted on a federal loan and was forced to shut its doors. After the mill closed, vandals and neglect took their toll on the buildings. In 1986, the property sold at auction for $8.5 million. John Beattie, one of Robert Beattie's descendants, was a partner in the group of investors that bought his family's mill.

That group, Affirmative Development, demolished 20 of the original structures to make way for the parking lot (which is on the street side, not the river side of the buildings), a swimming pool, a four-acre wetland and a terraced plaza that overlooks the falls. Affirmative converted the remaining riverfront buildings into condos whose 12-foot-high windows look out on the Passaic. It really is a lovely complex, one of the few riverfront development projects that actually treats the Passaic as a scenic plus.

After leaving the boats in the parking lot, we wander down to The Mill's pretty plaza. A waist-high green iron railing rims the plaza's river side. There are wooden benches and café tables and chairs and a bronze plaque to Robert Beattie. The inscription assures us that The Mill at Little Falls is "Dedicated to preserving the ecology of the river for the future."

Of course, the ecology of Little Falls had been drastically altered before the owners of The Mill came along. The terrace we are standing on actually overlooks a manmade dam that was erected above the original falls. The dam is one of several, largely unsuccessful efforts at Passaic River flood control. There were once two waterfalls at this site. They were oriented perpendicular to one another. The upstream falls ran parallel to the current. The river swept over a rock wall that was 10 feet high and 300 feet long. After sheeting over this long, shallow upstream ledge, the river pinched into a narrow chute— barely 25 yards across— and came thundering over the second ledge, which cut straight across the channel, perpendicular to the flow.

The downstream falls was 16 feet high, nearly twice as steep as the first, much wider cascade. By the time the Passaic made it over both ledges, a journey of less than half a mile, the river had narrowed by two thirds and lost more than 50 feet in elevation. The slender furrow in the Watchungs at Little Falls was and still is a major cause of Passaic River flooding.

The original waterfalls were dynamited into oblivion sometime after 1886 in an effort to widen the channel and mitigate the flooding. There isn't a soul alive today who has seen the twin cascade, and there are no photographs of it. However, Little Falls was a favorite subject of America's 19th Century landscape painters, and several did manage to capture the beauty of the two cataracts before they were destroyed. One of the last renderings of Little Falls is an 1886 oil painting by the Belgian artist William C.A. Frerichs[3]. The painting hangs above a small blue settee in the waiting room of the Manor House at Ringwood State Park in Ringwood, a woodsy borough in northwestern New Jersey where my sister lived for about ten years when she was first married.

My paddling companions and I are busy snapping pictures of Little Falls and admiring the view from The Mill's terrace when a young security guard strolls up. Several condo residents have apparently been calling him to ask who we are and why we're photographing the falls. He asks us politely but firmly to leave. This won't be the last time we engender suspicion on the trip. We consider defending ourselves, but resist the temptation. We should get going anyway. We gather up our gear and head back to the parking lot where we left the boats.

The put-in that Carl scouted for us is several hundred yards downstream from The Mill. Confirming that our wheelies are still secure, we roll the boats out of the condo parking lot and turn left onto Main Street. We follow this central artery east across Union Avenue. We are the only pedestrians strolling this section of Main Street. Drivers and passengers in the passing cars seem to take no notice of our little caravan.

---

3 William C.A. Frerichs came to the United States from Belgium in 1850. He is considered a minor member of the Hudson River school, a group of landscape painters that included Albert Beirstadt. In 1855, Frerichs accepted a teaching position in Greensboro, North Carolina. He lived there for 10 years, and is best known for his large landscapes of the Blue Ridge Mountains. Frerichs moved back north in 1865 and settled in Newark. Twenty-one years later, he painted the oil portrait of Little Falls that now hangs in the Ringwood Manor house.

This is my first urban portage, and it's going pretty well so far. Compared to what Lewis and Clark went through, portaging around Little Falls is a breeze. A few months earlier, in August, I had taken a canoe trip down the Missouri River in Montana. During that trip, I visited the Lewis and Clark Interpretative Center in Great Falls. There's a life-size diorama there: a half dozen men in buckskins strain to drag a very large wooden boat up and over a steep forested crag. You have to admire that pluck. Then again, Lewis and Clark never portaged a Pungo through 21st Century New Jersey. The sidewalks in Little Falls— and, I would come to learn, in Paterson and Garfield as well— are not exactly boat-friendly. Once we haul our kayaks across the busy Union Avenue intersection, we still have to get all 14-plus feet of plastic off the street, which entails getting it up onto a sidewalk and suffice it to say, there's not a sidewalk in Little Falls— indeed, precious few in all of New Jersey— wide enough for that.

It's striking how little boat access there is on the Passaic. This is true most everywhere along the river's 86-mile length, but especially at Little Falls and, as I would soon learn, at Great Falls, two of the most scenic and historic sites on the river. There is simply no safe place to get a boat into or out of the water at either location. Forget anything as civilized as a boat launch; there isn't even an accessible, public stretch of bank to be found. Which brings to mind something Andy Willner told me on the day we met. We were looking at the river from the parking lot of a strip mall on River Road in Harrison, New Jersey. Harrison is some 20 miles downstream from here, just north of the river's mouth in Newark. Andy was explaining that commercial developers who build along the river are required by law to provide some kind of public access to the water. It would be nice if every riverside town and city were required to provide a little public access too. Alas, they are not and do not and so getting our kayaks in and out of the Passaic is a logistical and physical challenge.

When Carl was scouting put-in and take-out points for this weekend's paddle he weighed two factors: one was the relative ease of getting the kayaks in and out of the water, the other was the distance we'd have to haul them between exit and re-entry points. Sometimes, he had to compromise.

A week or so before our October kayak trip, Carl emailed me some last-minute details, including this brief description of our portage plan for Little Falls: "The Little Falls are quite beautiful but difficult to get around," he wrote. "The portage is a 2,000-foot walk and the put-in is a steep— but soft— embankment."

"Copy that, Rambo" is what I emailed blithely back. But now I am standing in the public park just off Main Street in Little Falls, looking down a steep, muddy rut in the hillside. The Passaic is below me. *Way* below me. This is the put-in.

Little Falls is basically a notch in the middle Watchung Range. The Passaic had to travel some 40 miles north before it found this little chink in the basalt fortress. You can almost feel the river's joy as it gambols through the ravine below us. Free at last! The banks of the river here are sheer rock walls about 40-feet high. We're standing on top of them now and we have to get the boats and the gear and ourselves down to the river below. Cathy and I exchange a quick look that tells me she's thinking what I'm thinking: "Holy shit."

Carl's plan is to belay the boats one at a time down the narrow chute before us. It's actually a muddy gash eroded into the steep hillside. Cathy goes first, picking her way down along the slick rocks and the tangle of exposed tree roots. I mark her route intently so I can duplicate every move when it's my turn. We're not the first ones to get down to the river this way. Someone has tied an old green garden hose to a tree about halfway down the chute. Cathy grabs hold of it and uses the hose to lower herself down the last, steepest part of the descent. Okay. It isn't as bad as it looks from up here.

My boat will be the first one down. Carl ties a rope to the stern and starts easing the orange kayak down the slope, bow first. The Pungo gets hung up on roots in a few places, and I have to slide down behind it and nudge the stern free with my foot. But it's a pretty smooth belay, and quick.

The riverbank below is beautiful, a perfect place to put-in. Cathy and I carry my boat about 30 feet along the cool, shady bank to the river's edge. We set the Pungo down beside the remains of an old campfire. High above us a narrow green steel bridge crosses the river. The bridge carried an old Morris Canal aqueduct across the Passaic. This is where Albert Handy and his companion would have carried their canoe from the river to the Canal. I sit on the cool shady bank while Cathy climbs back up the chute to help Carl with the double.

The riverbank here is covered with a thick scree of flat, fist-sized shards of slate that have sloughed off the sheer rock face behind me. Several yards upstream this rock wall juts out into the water, anchoring the aqueduct bridge on this bank. The river tumbles down through a field of boulders and rock outcroppings, but this rock wall, like a headland or a jetty, deflects the flow towards the center of the channel, creating a protected swirling pool in its lee. The current is swift in the center of the channel, but here behind the rock wall the water is shallow and calm.

After what seems like a long time I walk over to the bottom of the chute to see what's keeping Carl and Cathy. As I get close, I hear Carl yell out from above and then I hear a strange crashing sound, and then I see the blue boat—flying solo down the chute. And I mean flying.

The kayak skids and scrapes and grinds and bounces and ricochets off rocks and roots and tree trunks and catches air before coming to an abrupt halt—nose first and on its starboard side—at the bottom of the slope, its wild, bruising descent halted at last by a grey boulder and two saplings. Had those obstacles not been there, the boat's momentum may have carried it all the way to the river.

For a moment, no one moves. If the blue double is damaged it could mean the end of the trip, at least for today. Carl lopes down the chute looking horrified. We all gather round the boat, and stare down at it. If the boat were a body we'd be feeling for a pulse.

"The knot slipped," says Carl, finally. He shakes his head. He's flustered, which isn't like him, and clearly embarrassed about the knot. We right the kayak and begin inspecting it for holes or other damage, bending over to peer into the hatches and running our hands along its sides and bottom. Carl swears he will never use plastic rope again. He says this twice.

Lucky for us, hard plastic kayaks are apparently indestructible. The Perception suffers a three-inch scuff on its stern. A Phillips Head screwdriver has punctured one of the dry sacks that was aboard. But otherwise, the blue double is good to go. We exchange sighs of relief and launch the boats quickly while the river gods are still feeling beneficent.

One lazy push of the paddle takes me to the middle of the channel where the river grabs hold of my boat and whisks me swiftly downstream towards Paterson and Great Falls. Giddy with the current's power I dig my paddle in and pull hard, then harder still, until the banks on either side become a blur.

I learned how to row a boat in Andover. My father taught me on our blue aluminum skiff. He showed me how to grip the wooden oars and extend my arms forward and down, then dip the oar blades into the water behind me and pull back. Reach, dip, pull. Reach, dip, pull. It's all in the legs, he'd say.

I can feel the Passaic's current begin to flag about a half mile downstream from Little Falls. The river channel suddenly widens again. In quick succession Carl, Cathy and I pass under the small railroad span at Lackawanna Avenue in Little Falls, around a gentle bow in the river channel and under Route 80, which skirts the southern edge of Paterson. Our next pullout is at Libby's Lunch, a little riverside joint in Paterson that has been slinging diner fare from the same spot on McBride Avenue above Great Falls since 1936. Libby's is only about a mile and a half downstream from the Route 80 Bridge. We must be getting close by now.

We are watching for two large parks that flank the river. Pennington Park will be on the right or eastern bank, Westside Park on the left. Past the parks, the Passaic will bend gently to the left. Just after that jog, but before we reach the McBride Avenue Bridge, we'll see the Libby's parking lot on the right bank. That's where we want to pull off the river.

Paterson's two riverside parks are impossible to miss. Pennington comes first. Nearly half a mile long, it sweeps up from the river in a long grassy run. On a basketball court about halfway up the slope, eight teenage boys are showing some moves. Spinning, shoving, talking trash. Their baritone exchanges pulse down the slope towards us like machine gun fire. Suddenly, the pickup game stops. The players have noticed us out on the river. They fall silent for a moment. Then one of them, I can't tell which, lets out a full-throated howl of warning: "WaterFALL!"

8 GREAT FALLS

IMAGINE THAT YOU ARE A BIRD, a strong athletic raptor like an osprey or a peregrine falcon. You are flying low over the Passaic River, following the current north towards Paterson and Great Falls. Although the current is swift, the surface of the water is smooth beneath you. You course quickly and easily along this polished mirror, letting your wing tips brush its dark surface now and then, leaving pairs of perfect round ripples in your wake.

A low bridge ahead— the Wayne Avenue Bridge— tells you that the falls is near. You dip beneath the short span, then over the low dam just beyond. Past the dam the water's sleek perfection disintegrates as the river tumbles across a shallow, terraced shelf of rock. The shelf ends in a vertiginous ledge, 77 feet high and 280 feet in length. You have reached the very top of Great Falls.

This rocky shelf is a launch pad, the planed apex of a basalt bluff. The shelf's sharp edge parallels the current on your right, as if some force had sheared off the riverbank. The yawning breach invites the Passaic to spill out of its natural channel and over the side. The river does not hesitate— and neither do you.

You bank right with the current, tuck your wings and follow the river over the ledge. Straight down, down, down you fall, through a thick curtain of water and the roar of river punishing rock, into a dark, claustrophobic crevasse. The crevasse is a crack in the easternmost Watchung Mountain, not more than 20 feet wide at its narrowest. Some epic shudder must have rent the basalt ridge here. The rock face is cleaved in two and the riverbed below is littered with breakaway boulders bigger than Buicks. The upstream bluff, your take-off point, is the southern half of the split ridge. It is shaped like a backward "J." The soft curl at the bottom of the "J" cups its mate across the river, a jagged steeper U-shaped bluff. The Passaic swirls madly between this pair, careening south then north through the steep, dark alley of rock.

A quick glance to your left confirms that you have entered a blind canyon, a dead-end in the basalt maze. You wheel right, towards the mouth of the crevasse and freedom, skimming the foam as you and the river follow the fracture line south between the two rock faces. At the exit, you pivot left with the current and hairpin around and between the two great bluffs. Now you and the river are moving north again into a flatter, gentler landscape. Soon, the thrill ride will be over, the thunder of the falls fading with every wing beat.

Great Falls is the Passaic's final spectacular gesture, a daring muscular stunt. WaterFALL! Once the river crashes through the final Watchung gorge, it flows wider, flatter and mostly straighter through its Lower Valley, the choked hallway that hosts the Passaic for its final 20-mile sprint to the sea.

Paterson and Newark bookend the Lower Valley. Newark, New Jersey's largest city, rises at the end of the Valley from the great, lost salt marsh that once greeted the river as it washed into Newark Bay. Paterson, the state's third-largest city, commands the head of the Valley from this ridge in the Watchung Mountains. The Passaic is the watery thread that links and animates these two Lower Valley titans. In Newark, the placid Passaic provided easy access for trade. In Paterson, the cascading Passaic provided power for industry. For better and worse, the fortunes of the two great cities and the river that connects them are entwined.

Paterson was founded in 1792, two years after Washington, D.C. and more than a century after Newark, its larger downstream neighbor. It was founded for one purpose: to make things. Paterson was America's first planned manufacturing city. As such, it owes its existence and all of its economic success to the Great Falls of the Passaic River. The story of how Great Falls gave birth to Paterson is epic and cautionary. It is a testament to America's bold entrepreneurial spirit, and a case study in how industrialization warped the way we saw, spoke and thought about the country's natural resources.

The environmental destruction of the New World predated its industrialization, to be sure. The founders of Newark had been mangling salt marshes for 100 years by the time Paterson was incorporated. America's earliest settlers felled whole forests in New England as they cleared land for crops and towns and harvested trees for export. (White pines were especially popular with the British Navy, which turned the tall straight trunks into ship masts.) In her book *American Environmental History: An Introduction*, environmental historian Carolyn Merchant notes that by the early 18th century hundreds of acres of New England timber were already gone, and populations of native birds and mammals were drastically depleted. Even so, the beauty and wildness of nature were still much in evidence and duly valued in pre-industrial America. Back then, nature was appreciated as a wonder of God.

Attitudes about natural resources began to change with the emergence of an American manufacturing economy. Once nature became a commodity, reverence for the wild and all its denizens began to fade. Forests were logged, rivers dammed, marshes drained, animals slaughtered for pelts and feathers. With swift efficiency, the wild places were tamed, and in the taming their beauty and wonder were largely lost. America's natural resources, its mountains, rivers and forests, were seen first and foremost as raw materials to be harvested for the production of goods. The wonder of God was fast becoming an economic instrument of man.

Where painters and poets once labored to capture the beauty of Great Falls, engineers and entrepreneurs now began to quantify its assets. Suddenly, the value of the Passaic at Great Falls was being measured in horsepower: how many mill wheels could the river's surging waters turn, and how fast? The carved chasm, the thundering spill, the rainbow streaked mist that filled the air. None of these scenic attributes could be assigned a production value, and so they were not counted. The tension between production and preservation had begun, and preservation was about to get its ass kicked.

"Living nature," writes Carolyn Merchant, "disappeared from the everyday experience of most Americans by the mid-twentieth century." The Great Falls of the Passaic River was a pivot on which the nature paradigm shifted. Legions of skilled tradesmen were already churning out goods in Newark. The emergence of Paterson, a planned manufacturing powerhouse, cemented New Jersey's reputation as a vital industrial hub, and sealed the Passaic River's fate.

The notion to use the power of Great Falls to turn the wheels of progress was the brainchild of Alexander Hamilton, the man on the $10 bill. Hamilton served in the New York State legislature and represented New York at the Continental Congress before becoming the first U.S. Secretary of the Treasury in 1789. One fanciful tale has Hamilton first encountering Great Falls in July 1778 while he was serving as a Colonel in the Continental Army and a senior *aide-de-camp* to General George Washington. Hamilton and Washington were said to be journeying to Paramus with a pair of French officers when they stopped to picnic on a bluff overlooking the falls.

"It's poetic," said Paterson historian Nick Sunday about the picnic story. "But it's conjecture." According to Sunday, there is no "direct evidence" that Hamilton ever picnicked above Great Falls with his fellow officers. "But it's not important," he added. "What matters is that Alexander Hamilton saw the falls at some point."

Today, Great Falls is surrounded and almost hidden by the city of Paterson. Whenever he first beheld it, Alexander Hamilton would have seen the falls in its wild and unobstructed state. It must have been magnificent. The Paterson falls has been called a miniature version of Victoria Falls, the world's largest cascade. Victoria Falls, on Africa's Zambezi River[1], is certainly larger— 329 feet high and more than a mile long. But the zigzag courses that both rivers cut through the gorges below their falls are similar. The Passaic reverses direction twice as it surges over and through the basalt canyon at Great Falls. From above, its whiplash track resembles the letter "N."

Hamilton, the soon-to-be treasury secretary, saw more than a breathtaking spectacle of water and rock. He envisioned Great Falls as the site of a manufacturing center that would help the new country rival, and eventually eclipse, Great Britain's industrial supremacy. England guarded its manufacturing superiority carefully. The crown controlled all investment

---

1 Below Victoria Falls, the Zambezi knifes back and forth across the landscape through a half dozen Z-shaped gorges in the basalt plateau. Each of these downstream gorges was once the deep chasm beneath the falls. Over time, the falls of the Zambezi River receded gradually upstream to its present-day drop at Victoria Falls, leaving the old gorges behind.

capital and dictated the rules of trade. American-made goods were not allowed to enter England, and England's machinery, industrial technologies and skilled laborers were not allowed to leave. When it came to manufacturing, the relationship between the crown and its premier colony was strictly one-way: America shipped raw materials to England, which sent manufactured goods back to America.

For Hamilton, Great Falls was a way to break this stranglehold on America's economic growth. He reasoned that a robust base of domestic manufacturing would generate much-needed capital for the new country and at the same time reduce America's reliance on foreign-made goods. Hamilton's plan for Great Falls was a cry of economic independence.

A decade later, as Treasury Secretary, Hamilton made his case for domestic manufacturing in a lengthy treatise titled *Report on the Subject of Manufactures*, which he submitted to Congress on December 5, 1791. In it, Hamilton listed seven advantages of a domestic manufacturing economy. The first was the "proper division of labour [sic]." American farmers could certainly go on making their own shoes in between harvests. But wouldn't farm families— and shoes— be better served if there were other workers dedicated specifically to this task? Specialization, wrote Hamilton, "causes each [occupation] to be carried to a much greater perfection." Specialization also saves time, that most precious commodity, "by avoiding the loss of it, incident to a frequent transition from one operation to another of a different nature."

Hamilton also extolled the use of machinery, which "has the effect of augmenting the *productive powers* of labour, and with them, the total mass of the produce or revenue of a Country." (The italics are his.) He predicted the creation of new and different kinds of jobs for "persons who would otherwise be idle (and in many cases a burthen [sic] on the community)." Finally, the Treasury Secretary noted that manufacturing would lure skilled immigrants "from Foreign Countries" and generate a more steady demand for goods.

The view of Great Falls that Alexander Hamilton outlined in his 1791 report was empirical and economic. He converted natural endowments into business assets, translating water volume and current speed into productivity and revenue. His argument introduced new concepts (specialization, switching costs) and a new language ("division of labour," "economy of time") that would enter the discourse and gradually transform the country's economy, ecology and psychology.

Hamilton's presentation and his vision for a domestic "manufactory" failed to stir the U.S. Congress, which lacked the will to create a second federal city. Nor did it please Thomas Jefferson, his cabinet colleague[2], who preferred that America remain an agrarian society. Jefferson warned that an embrace of manufacturing would weaken the young country's character and corrupt its very soul.

But Hamilton's idea did find purchase among the wealthy entrepreneurs in New York and New Jersey who grew intrigued by the profit potential and by the unassailable logic of Hamilton's proposal. What better place than Great Falls to test this new economic theory? It was close to the

---

2 Thomas Jefferson, the nation's third president, served as the first U.S. Secretary of State from 1789 to 1793, the period during which the Society for the Establishment of Useful Manufacturers was formed. A decade earlier, in 1781 while still the Governor of Virginia, Jefferson explained his opposition to manufacturing in the influential *Notes on the State of Virginia*. This collection of 23 "Queries" included Jefferson's views on everything from climate and population to laws, religion and military force. Query XIX, one of the shortest, dealt with the subject of "Manufactures." In it, Jefferson argued that America should not follow the British down the path of industrialization, but focus instead on cultivation and husbandry, which in Jefferson's view were more practical and principled occupations. "Those who labour in the earth are the chosen people of God, if ever he had a chosen people," wrote Jefferson. "...Corruption of morals in the mass of cultivators is a phenomenon of which no age nor nation has furnished an example." Woe to the country whose citizens come to "depend...on the casualties and caprice of customers," he continued, for that "dependence begets subservience and venality, suffocates the germ of virtue, and prepares fit tools for the designs of ambition."

major ports and population centers of Philadelphia and New York, and it offered ready access to the raw hydropower of the Passaic River.

Working quietly behind the scenes, and not in his official capacity as Treasury Secretary, Hamilton assembled an investment group of mostly affluent New Jerseyites. The group called itself the Society for the Establishment of Useful Manufacturers (or SUM). In 1791, the state of New Jersey awarded the SUM exclusive land and water rights to a six-acre area around Great Falls. The state's charter conveyed extraordinary powers to this new privately-held entity. The SUM received a permanent exemption from all local taxes and the freedom to alter the course of the Passaic River and its tributaries in any way it saw fit. It was "the first time in history that an organization was given the exclusive rights to harness an entire river," noted Paterson historian Nick Sunday. "Before that, rivers were owned by the king."

During the next 50 years, Hamilton's dream of a manufacturing stronghold would be wrought in brick and steel along the banks of the Passaic. The SUM began by acquiring 700 more acres of land above and below the falls. It constructed a few factories and mills along the river and a wooden dam across it. The 13-member board of directors hired architect and engineer Pierre Charles L'Enfant to create a system of raceways that would capture and distribute the river's power more efficiently. L'Enfant was also charged with designing a model manufacturing city at the edge of the falls.

The city was to be named Paterson in honor of William Paterson, the New Jersey Governor who signed the charter that established the SUM.[3]

Pierre L'Enfant had won much praise for his work on the design of the nation's new capitol in Washington, D.C., and he conceived Paterson as its sister city. Washington was the great nerve center of the new republic, enacting laws and making policy. Paterson would be its manufacturing muscle, turning out goods for domestic and foreign markets. L'Enfant's plan for Paterson included wide stately boulevards— to enhance airflow and aesthetics— and fountains below the falls. "It was a baroque masterpiece," said Nick Sunday. "It had panache."

But despite the initial support and investment, grand plans for the new city and its groundbreaking manufactory went unrealized in those early years. L'Enfant's ambitious—some called them grandiose—plans for Paterson, which involved an aqueduct to divert river water for SUM operations and the monumental urban design, were deemed impractical and too expensive. Within a year, SUM directors dismissed L'Enfant and brought in Connecticut state treasurer Peter Colt to supervise the project. Colt simplified his predecessor's original design and managed to finish the first raceway in 1794. But by then the SUM was languishing, a victim of dwindling cash reserves, a dearth of skilled labor, its own poor and, on occasion, unscrupulous management and, in the view of Nick Sunday at

3 Paterson historian Nick Sunday contends that the city's name actually honors the Scottish William Paterson (1658-1719), who after founding the Bank of England in 1694, attempted to start a trading colony called Darien on the Isthmus of Panama. Paterson's wife and son died during the failed expedition, and the Scottish government, which had underwritten the Darien scheme, lost its investment. In the wake of this ill-fated adventure, Paterson [illegible] the losses, and as a way to gain emigration rights to America. "[Alexander] Hamilton called the city Paterson, but everybody knew that it was named after the *other* William Paterson—considered the liberator of the Scots—for whom the Governor of New Jersey was probably named."

least, the untimely inattentiveness of its greatest champion. "Hamilton was having an affair with a married woman from Morristown, and her husband pretty much blackmailed him," said Sunday. "It created a tremendous distraction." Things got so bad for the SUM that in the spring of 1796 it was forced to shut down operations altogether and offer up its properties for rent. These dire circumstances might have spelled the end of Hamilton's grand vision had it not been for two other Colts: John and Roswell.

Roswell Colt was the son of SUM supervisor Peter Colt. John Colt, no relation, was a local engineer. Along with a few like-minded Patersonians, the two business partners purchased a majority interest in the SUM and turned the failing Society for the Establishment of Useful Manufacturers into a successful real estate development venture.

Under its new management, the SUM immediately began leasing its existing factories and warehouse space along with the land and water rights that went with them. Meantime, John Colt began engineering a better dam and a more efficient three-tiered raceway system. The work was finally completed in 1838. Colt's clever improvements, which remain in place today, enabled the SUM to more carefully siphon the Passaic.

With Colt's fine-tuning, the SUM could now develop new tracts of land along the new raceways and sell these new river-powered parcels to industrialists eager to build new mills. By upgrading infrastructure, selling properties in its new industrial subdivisions and leasing the water rights, the SUM began to flourish. And so did the city of Paterson. In 1946, the city purchased the SUM for $450,000. By that time, the Society's holdings also included a hydroelectric power station at the foot of Great Falls and a steam generating plant nearby, which was used as a backup to supply power when the river was running low.

Although events did not unfold in exactly the way that Alexander Hamilton had intended, Paterson did become a manufacturing center.

Skilled laborers from around the world came to work in its mills and factories. Like Newark, downriver, 19th Century Paterson laid claim to many industrial firsts:

The Wright Aeronautical Corporation of Paterson built the "single, air-cooled, Wright Whirlwind J-5C nine cylinder radial engine" that powered Charles Lindbergh's "Spirit of St. Louis" nonstop from New York to Paris in May 1927. By 1929, the Wright plant was turning out more than 6,000 engines a year.

Machinery designer Thomas Rogers launched the first of Paterson's five locomotive manufacturers in 1835. Within 40 years Paterson was producing nearly 80 percent of all America's steam locomotives.

In 1836, while Newark's Samuel P. Smith was wowing woodworkers with his clear non-stick varnish, Samuel Colt was producing the "Paterson pistol," America's first repeating revolver, in his gun mill at the corner of Mill and Van Houten Streets. (Samuel Colt was the nephew of Peter Colt.)

That same year John Ryle, a former British bobbin boy, developed a method for spinning silk on spools. The invention would make Ryle the country's most successful silk merchant and earn Paterson the nickname "Silk City." By 1870, the year electrical inventor Edward Weston arrived in Newark, Paterson was producing half the nation's silk. At its peak in the early 20th Century, the city's silk industry employed nearly 15,000 people.

In May 1878, Irish-born inventor John P. Holland piloted his new one-man submersible on a test run in the Passaic River near Paterson's Spruce Street Bridge. The sub sank. But Holland survived and persisted. He is known today as the "father of the modern submarine." His original 14-foot sub, the *Holland Boat No. I*, was salvaged from the Passaic in 1928 and resides today in the Paterson Museum.

Paterson was also a major producer of cotton, wool, flax, hemp and all kinds of apparel. In fact, the city's contributions to manufacturing were so significant that in 1977 The American Society of Civil Engineers and the American Society of Mechanical Engineers declared the "Great Falls Raceway and Power System in Paterson, N.J." a National Historic Mechanical and Civil Engineering Landmark. Program notes from the dedication ceremony on May 20th of that year describe the Great Falls complex as "the basis of the oldest American community integrating water power, industrial development and urban planning."

The engineering marvels that enabled the SUM to convert Passaic River power into guns and silk, locomotives and airplane engines guaranteed Paterson's legacy as "The Cradle of American Industry." The Great Falls area has been a National Historic Landmark since 1976. In 2004, it became a New Jersey State Park. When, in April 2009, President Barak Obama signed long-pending legislation inducting Great Falls into the National Park system, the 35-acre site in downtown Paterson took its place alongside Yosemite and Yellowstone and Arizona's Grand Canyon as one of America's greatest scenic treasures.

Of course, the explosion of manufacturing along the banks of the Passaic took its toll—on the river, the city and the falls. As Paterson's mills and factories and population expanded, industrial and human wastes polluted the Passaic. Paterson never developed a chemical industry, and so it was spared Newark's dioxin and PCB problems. But heavy metals were in heavy use by the city's many dye shops. When the river water in Paterson grew too dirty for dyeing fabrics, the city was forced to pipe cleaner water from upstream down to the dyehouses. The old "dyers line" is still visible, a green pipe across the falls. Today, there are more than 70 superfund sites in Paterson, many of them at former dyehouse sites along or near the Passaic.

In 1902, a great fire swept through Paterson. The very next year, a massive Passaic River flood inundated the city. The back-to-back disasters damaged or destroyed many of the city's mills and factories and, in Nick Sunday's view, marked the beginning of a long slow decline. "The heyday for Paterson was 1901," he said. "And it was a rotten heyday, because the underpinnings of the collapse were all in place."

In the early years of the 20th Century, Paterson's silk industry was rocked by chronic labor unrest and threatened by growing competition from the new, cheaper synthetic fabrics. Friction between Paterson's silk merchants and their immigrant workforce spawned nearly 140 strikes between 1881 and 1900. Tensions boiled over in 1913 when thousands of silk workers walked off the job in a bid for better wages and shorter workdays. Paterson's silk factory workers typically clocked five, 10-hour days each week, plus half a day on Saturday, and their pay was well below industry average. The 1913 strike lasted seven months and ended in defeat for the workers. In its wake, mill owners began moving their businesses out of Paterson.

The great silk exodus, mostly to Pennsylvania where labor laws favored management, set the stage for an erosion that plagues the city of Paterson to this day. Nick Sunday blames his hometown's decline on the utter lack of vision from its leaders. "They weren't administering for the future," he said. "They were losing textile manufacturing. Social institutions were not really in place. The schools were not very good. Illiteracy was high. All the things the city needed to survive bad times weren't here. The mill owners weren't thinking about society and building institutions that would last. The politicians were parasites. There was corruption. You didn't have visionaries, people who could see past the present. They didn't prepare the city for the onslaught of the Depression, etc."

Small and mid-sized industries still operate in Paterson today. But the city is not the industrial powerhouse it once was. Nor did it ever achieve the kind of grandeur that Alexander Hamilton and Pierre L'Enfant, his hand-picked urban architect, envisioned. The missed opportunities pain Nick Sunday. "Paterson wasn't supposed to be just another New Jersey town," he said. "It was supposed to be a grand place. Grander than Washington, D.C. A place where people would come to improve themselves, not be thrown into poverty because the industry left the city. The magnificent metropolis that L 'Enfant and Hamilton talked about didn't materialize. The silk works came in and took over. They narrowed all the streets, put in big buildings for their mills. They abandoned L'Enfant's plan. They made it a work-a-day town." To their credit, Paterson's silk merchants employed thousands of workers and left the city an impressive art collection, which is housed in the city's Public Library. But to Nick Sunday, historian and native son, "silk really ruined Paterson."

The same could be said for the effect of the SUM on Great Falls. Silk was a water-intensive business, especially the dyeing part. As the Society for the Establishment of Useful Manufacturers exercised the full measure of its control over the Passaic, it diverted more and more flow for more and more mills and factories. At peak production times, especially during the river's low-flow summer months, all that remained of Great Falls was a trickle of river dribbling over the lip. There was no more thunder, no mist, no rainbows. The power that produced them was busy elsewhere, turning mill wheels and drive shafts and factory belts. When the Society siphoned off the Passaic's water, it siphoned off the splendor of Great Falls too. The sad image of the "dry falls" crept into the art and literature of the day.

Pomona, a character in the 1879 novel *Rudder Grange*, described her honeymoon visit to Great Falls like this: "So we thought of Passaic Falls, up to Paterson; An' we went there…and walked over to the falls. But they wasn't no good, after all, for there wasn't no water runnin' over 'em. There was rocks, an' precipers, an' direful depths, an' everything for a good falls, except water, an' that was all been used by the mills. Well, Miguel, I says, this about as nice a place for a falls as I ever see…"

Tourists like the fictional Pomona used to flock to the falls. But visitors showed little interest in its new water-starved state. Robbed of its former might and much of its beauty, the falls lost the power to lure and wow them. It just didn't seem that special anymore. In this way, Great Falls suffered the fate of so many Passaic River marvels: it was commandeered and degraded, then ignored and forgotten. I grew up 20 miles from Paterson. My mother went to nursing school there. Neither of us had ever been to Great Falls. I had no idea it even existed.

My mother and I made the pilgrimage to Paterson one chilly Sunday morning in April 2003. My mother would die in October of that year. But of course we didn't know that as we stood on a high footbridge overlooking Great Falls. I remember being stunned in equal measure by the majesty and the mess. I was expecting some wear and tear and urban grime. But I was unprepared for the extent of the abuse and the overall neglect.

I didn't think it was possible to spoil such an impressive natural wonder, but the area surrounding the 77-foot-high cascade was ravaged. Weeds choked every pathway. Broken glass littered the parking lot. Graffiti scarred the basalt cliffs. Scraps of garbage floated in John Colt's raceways. Piles of it collected at the edge of the dam and in all the little crevices and backwater pools above the falls. It was your basic gumbo of disposable bottles and cans, cups, wrappers and fast food containers, the handheld trash that's easy to deposit in a garbage can, but even easier to toss out a car window or abandon in a parking lot or gutter. This monument to human carelessness reminded me of a Great Falls blog entry I had stumbled across in my research. "My impression of the Great Falls from the first time I saw them was: except for those waterfalls, this place is damned ugly," wrote Bob, the blogger. "Alexander Hamilton's America isn't pretty."

My mother didn't seem troubled by any of this. She was too dazzled by the falls. I whipped out my new video camera and with the cascading Passaic as a backdrop, taped her talking about her nursing school days. How lots of registered nurses were serving overseas at the time, which meant hospitals were forced to use nursing students like my mother on the floors. She remembered being terrified that she'd make some awful mistake. But she did well. She graduated in 1944, first in her class.

A respectable volume of river was coming over the cliff that April morning. You can barely hear her voice above the roar.

Carl, Cathy and I take the boats safely out of the Passaic at Libby's, where we'll stop for a late lunch. We drag the Pungo and the Perception over dried tufts of grass, shattered glass and assorted trash up a low raked bank past a thick poison ivy-entwined tree. Carl clambers easily over the two barriers at the top of the bank—a metal post fence and the hurricane fence behind it cordon off Libby's large asphalt parking lot. One by one, Cathy and I hand him the two kayaks, then climb over after them.

We slip on the wheelies and are just beginning our march across the lot when a shiny black SUV with dark tinted windows cuts across our path and stops. I gird myself for a scolding, or worse. But when the passenger window purrs down, the 20-something couple inside is all smiles and wonder and rabid curiosity.

Had we come down the river? They want to know. What was that like? What kind of boats are these? Where are we going? Do we know about the falls? How will we get around them?

I believe they would have kept on asking us questions but for the car that pulls up behind them, its driver clearly impatient to exit the lot. The SUV couple wishes us luck, expresses once again their delight at our adventure, and rolls off, all smiles and waves and good cheer.

It's 4pm. The sun has disappeared behind a bank of dark clouds and the wind has begun to gust. We leave the boats over near the employees' entrance and duck into Libby's. It feels good to be indoors where it's warm and dry. The place doesn't really qualify as a diner, not by Jersey standards anyway; the menu is too limited. It's more like a hotdog stand with cocktails. In fact, beneath the "Libby's stands for *Quality* FOOD" sign that hangs above the door, you can still make out traces of faded blue text that once advertised Libby's "Hot Texas Weiners." Hot dogs all the way are the house specialty.

Even at this in-between hour on a Saturday afternoon, Libby's is crowded. It's a pretty tiny place, but still. The room is L-shaped. We walk down the long part of the "L," past the cashier and the bar on our left and a row of booths and windows on the right. The windows above the booths look out on McBride Avenue. Even though Libby's is right on the river, there are no windows that overlook the Passaic.

We slide into a booth near the back. There's a big round banquette behind us where two young mothers huddle over coffee while their brood of six small children competes for French fries. Our lunch arrives on paper plates. The food isn't fancy, but everything is fresh and simply prepared. And there's booze! If I weren't about to haul my Pungo back into the

Passaic and paddle for another five miles, I'd be sipping scotch. I settle for hot tea instead, and a charbroiled chicken sandwich.

I first heard about Libby's from Ella Filippone. This was back in September 2005, during my second visit to Great Falls. Ella had offered to walk me though Mary Ellen Kramer Park, one of several green spaces above the falls. The late Mary Ellen Kramer[4], former Paterson First Lady and fan of Great Falls, was a good friend of Ella's.

Ella is the current and founding Executive Director of the Passaic River Coalition, the oldest Passaic River advocacy group. She has survived 40-plus years of Passaic River battles, and won more than her share. She has authored reports, haunted public hearings and testified before Congressional committees, all in an effort to protect and revive the river and its watershed. She is the doyenne of Passaic River advocates.

We agreed to meet in Overlook Park beneath the life-size statue of Alexander Hamilton. His bronze likeness stands on the high J-shaped bluff just downstream from the falls. From this vantage point, Hamilton has a commanding view of the canyon and the cascade and of the green steel footbridge above it where I stood with my mother on that April morning. Down below is the red brick hydroelectric plant, which is operating again. A few feet from the statue there's a peeling blue metal sign that explains how "Alexander Hamilton envisioned the great potential power of these scenic falls for industrial development."

The mastermind of U.S. manufacturing looks out sternly over Great Falls. He seems troubled by what he sees. The SUM is gone, though

4 Mary Ellen Kramer was a Paterson native and the city's First Lady for eight years (1967-1971 and 1975-1979). When a state-proposed highway project threatened to destroy some of the city's old silk mills, her spirited opposition saved the buildings and led to their historic designation. Kramer continued her preservation work as Executive Director of the Great Falls Development Corporation. In 1980, she became Director of Community Affairs at William Paterson College. She died of cancer 13 years later. She was 53.

descendants of its original members still meet once a year to collect the interest from old investments. Hamilton's model industrial city has long since lost its status as a manufacturing center and is struggling economically. There is some good news these days. A new retail mall is going up next door to the Paterson courthouse, and the city and state have launched a multimillion project to refurbish a 42-acre area around Great Falls. Now that Congress has granted Great Falls its National Park status, federal dollars may also flow into the area. The long-awaited makeover of Great Falls promises to include new landscaping and walkways, restoration of the wrought iron fencing in Mary Ellen Kramer Park, and a refurbishing of the old steam plant. The National Park Service is reviewing more ambitious proposals, such as an interpretive center and a 500-seat amphitheater. But as I stood there waiting for Ella that September morning, I couldn't help imagining Hamilton's disappointment at what's become of his manufacturing dream.

Today, the falls is swaddled in 118 acres of neglected urban parkland. Great Falls Park, as it is collectively known, is a long thin green belt that twists north to south, mostly following the river as it winds through the southwestern portion of downtown. The cluster of surviving SUM structures along Mill and Van Houten streets is clearly visible off to the east behind the veneer of trees. This is the Great Falls/SUM Historic District, a red brick city-within-the-city. Rogers Locomotive, Cooke Locomotive, the Barbour Flax Spinning Company, the Casper Silk Mill, the Harmony and Industry Mills, the Dolphin Mill, Hamilton Mill, Franklin Mill, Essex Mill, Congdon Mill, Phoenix Mill and the crumbling remains of Allied Textile Printers, the complex that once housed Samuel Colt's gun factory and John Ryle's first silk operation. You can almost hear the hum of bygone machinery.

Some of the old SUM buildings still host industrial tenants—APS Contractors Equipment & Leasing, Phoenix Renewable Energy, Fabricolor, Inc. The original Phoenix and Essex mills have been converted into office space and artist's lofts. The old Thomas Rogers Locomotive plant houses the Paterson Museum and some offices. But many of the SUM buildings are empty now. From Overlook Park, where I waited, I could look straight ahead into the basalt canyon of Great Falls or off to the right for a glimpse of the old red brick facades.

Ella arrived a little after 10am. She appeared perfectly round and soft as she strode towards me, her face and body like the circles in a child's drawing. Her short hair accentuated the curve. Straight and fine, it lay flat against her cheeks like silver parentheses. But the gentle contours belie the steeliness beneath.

Ella's manner is commanding, even imperious at times. Her gift for narrative is impressive. Precise details may elude her here and there—it has been 40 years, after all— but the story arc is clear and concise. She navigates its twists and turns with ease. She knows where all the bodies are buried.

Ella imagines a rejuvenated Passaic, with pleasure boats plying its clean waters, well-tended greenery lining its shores, and a hand's off approach to flooding that lets the river reclaim its natural floodplain. Not everyone subscribes to Ella's vision for restoring the river. Naturalists like Carl prefer wilder solutions, replanting marshes rather than building the manicured parks she favors. Rogues like Andy Willner, who insist that polluters pay for Passaic River cleanup, have questioned Ella's coziness with the corporate community. (The PRC has accepted grant money from some of the Passaic River's corporate polluters.) But no one doubts her commitment to the cause: Ella wants the river back. "I'd like to get more rowers on the river," she told me. "More restaurants. It'll be nice to have riverboats again. You can do so much to enjoy the river, but it's been so encroached upon. It's been near dead. Last rites were said over it several times. I would like to see the Passaic have a life of its own again."

Ella grew up in Lyndhurst, New Jersey, just one town north of North Arlington. The Passaic forms the western boundary of our hometowns. She has rosier memories of the river than I do, largely because most of the Lyndhurst riverside is a Bergen County park. The riverbank there is a flat, green skirt with groves of locust trees, playgrounds, tennis and basketball courts and the baseball diamonds where my softball team practiced every summer. "I'd go down to the river a lot with my friends and hang out," said Ella. "There was a nice path and some big rocks you could sit on and we would just talk and have fun. That was our Sunday afternoon. Years later, I married Joe Filippone and moved up to Basking Ridge. And low and behold there was a Passaic River up there just like the one I knew where I grew up. Lots of people don't know how this river winds and turns all over the place."

Not long after she arrived in Basking Ridge Ella joined a local group called Citizens for Conservation. It was a loosely organized bunch of neighbors who convened in kitchens and rec rooms. Their mission was to protect the area from developers and industry, and from the agency that would become their chief nemesis: the U.S. Army Corps of Engineers.

When Congress created the Army Corps of Engineers in 1935, it put Passaic River flood management high on the new agency's agenda. A massive flood the very next year only underscored the urgency of the problem. During one eight-day stretch in March of 1936, a semi-tropical storm system dumped six inches of rain on the Passaic's already snow-covered watershed. The downpour washed away the snow, and the one-two punch of drenching rain and rapid snowmelt sent the Passaic surging over its banks. At the storm's peak, 19,400 cubic feet of river water—more than

145,000 gallons of Passaic—was gushing over Great Falls every second. That was the sixth worst Passaic River flood.

The mother of all Passaic River floods remains the deluge of 1903 when the swollen river poured over Great Falls at a rate of 37,000 cubic feet—or 254,338 gallons—per second. That's like emptying an Olympic-sized swimming pool of water over the falls every three seconds. No other Passaic River flood, not even the 2011 surge that followed Hurricane Irene, has come close to that peak discharge.

The storm that caused the 1903 disaster blew in on October 7th and lingered for four days. The Passaic was still at flood stage a full week after the storm moved out. Photos taken at the height of the flooding show foaming white water nearly filling the gorge below Great Falls to its brim.

Weather conditions that summer and fall had warned of trouble. The three months leading up to the October '03 storm had been extremely wet ones. The ground was completely saturated. When the big storm hit, the rainwater, and there was lots of it— 15.5 inches in five days— simply had no place to go.

Paterson got the worst of it: 11.45 inches of rain in one 24-hour period. Most of the precipitation fell through the night of Oct. 8th and into the following morning. The *Paterson Evening News* reported that the Passaic rose six inches every hour that night. By morning, downtown Paterson was underwater. Most of the city's bridges had been washed away. The Straight Street Bridge was done in by a floating barn. Miraculously, only three people died, all victims of drowning. One young girl, who was standing on the Haledon Bridge, disappeared beneath a wall of water that swept her and the bridge straight away. The flood left thousands homeless and more than $7 million dollars worth of damage, much of it to the SUM mills.

The river's flood record is sobering to say the least. It reminds me that although we have stripped New Jersey's nature nearly bare, we have not destroyed it. The Passaic River still has the power to strike back. The late author Kurt Vonnegut, in the wake of Hurricane Katrina, declared that "we are killing this planet as a life-support system," and that in response, "the planet's immune system is trying to get rid of us." Vonnegut was referring to the extreme weather and also to the emergence of deadly viruses such as HIV and the new resistant strains of flu and tuberculosis. But his observation recalled my childhood fear of the Passaic's revenge. Flooding was payback for all the times and all the ways we have tried our best to kill the river.

Disasters like the great flood of '03 were on the minds of Army Corps engineers as they conducted their meticulous studies of the Passaic River watershed. They presented their flood control recommendations in 1968, a year whose spring flooding had approached 1936 levels. The Corps' eventual proposal, which became known as Plan III, called for building two large dams, a large reservoir and a dry detention basin that could be used to hold excess storm runoff. These are the dams that would have permanently flooded the Millington Gorge.

Residents from upper Passaic towns—Basking Ridge and Bernards Township and other communities that border Millington Gorge—were not enthusiastic. They recognized the flood risk, but they were loath to see the Gorge and other parts of the Passaic's Central Basin sacrificed to a flood control plan that seemed, to them at least, like overkill. "It would have been a big dam and hundreds of miles of levees and flood walls and the wetlands flooded out," said Ella. "It was a really big project."

Ella and her fellow Citizens for Conservation took up the fight. The Corps eventually shelved Plan III and introduced an alternative called Plan IIB, a plan quipped Ella, which "was not *to be*."

With the Passaic now its major focus, Citizens for Conservation morphed into the Passaic River Coalition. The PRC incorporated in 1971. Five years later, Ella testified before the Congressional subcommittee that had convened hearings on the Army Corps' proposal. It wasn't that the citizens of north Jersey didn't want a Passaic River flood control plan, she told committee members: "We just don't want this one."

The PRC favored "nonstructural solutions" to flooding, such as offering buyouts to flood zone residents and restoring the original floodplains and wetlands that are the Passaic's natural defenses against high water. Their arguments were persuasive enough to deep-six both Plan III and Plan IIB and send Army Corps engineers back to their drawing boards.

By the mid-80s the Corps returned with an alternate Passaic River flood control plan. This one involved digging a 40-foot-wide diversion tunnel to collect Passaic River floodwater from low lying areas around Wayne and Pompton Lakes and pipeline the water to Newark Bay, a journey of more than 20 miles at a (mid-'80s) price tag of $1.2 billion. Alarmed, the PRC took on the Corps again. "The flood tunnel would have meant 17-foot-high flood walls in Lyndhurst," said Ella. "They were going to put up these berms between the parks and the riverbank. It would truly have destroyed the view of the river as you and I know it."

This time around it was towns along the lower Passaic that joined the PRC in its fight against the "tunnel monster." The flood tunnel threat got the attention of the American Rivers association, which put the Passaic on its Most Endangered Rivers' list in 1990 and 1991. The PRC gained another powerful ally in 1990 when Christine Todd Whitman came out against the flood tunnel during her Senate race against Democratic incumbent Bill Bradley. Whitman lost to Bradley, but three years later, in 1994, she won the New Jersey Governor's race.[5] "When she became governor," said Ella, "she told the Army Corps there was no public interest in the flood tunnel and the flood tunnel was dead and gone." But Passaic River flooding was not.

The PRC continued to push for alternative solutions, pressing its nonstructural case in Trenton and in Washington, D.C. Whitman remained an ally. But Ella credits former State Assemblywoman Maureen B. Ogden with finally prying loose the funds for a buy-back program.

Ogden was representing the New Jersey township of Millburn in the early '90s when the State Assembly held hearings on a buy-back program for flood-prone properties along the Passaic. "There had been a lot of storms," recalled Ella. "People on the coast wanted the state to buy out some houses, and Maureen Ogden said, 'Well, you're not going to pass a bill for acquisition of houses on the coast without one for the Passaic.'"

The Army Corps of Engineers now operates that floodplain buy-back program. Known as Blue Acres, it acquires homes and wetlands in the Passaic River Basin. At present, the Corps has purchased about 3,000 acres of flood storage land, including the Wildlife Preserves, Inc. property in Great Piece Meadows, and a handful of houses in soggy Wayne, New Jersey.

When the PRC first proposed purchasing private floodplain properties as a solution to Passaic River flooding, "many public officials scoffed at it," said Ella. "Now, an overwhelming number of people want to be bought

---

5 Christine Todd Whitman became the 50th Governor of New Jersey. She was the first and, as of this writing, the only female governor of the state. In 2001, President George W. Bush tapped Whitman to head the U.S. Environmental Protection Agency. She served for two years, resigning in June 2003 after several public disagreements with the administration over environmental policy.

out." Indeed, enthusiasm for buy-outs has skyrocketed in the wake of the August 2011 flooding. "The Blue Acres Program has been inundated with requests," said Ella. The PRC has been helping out with the purchase of 100 homes in Little Falls. Unfortunately, according to a recent report commissioned by New Jersey Governor Chris Christie, the $30-odd million in the state's current buy-back fund is a drop in the bucket compared to the projected $3.4 billion needed to purchase all the threatened properties in the Passaic's 10-year floodplain.[6] The state is also exploring other mitigation options, including the possibility of helping affected homeowners pay to elevate their homes above the reach of the river's floodwaters.

One hot summer night when I was about 10 years old I had a dream about trains. It was a classic sequence: two black steam engines billowing white smoke barreled towards each other down the same track. When they collided—which isn't supposed to happen in a dream, is it?—the shriek of brakes and twisting metal and exploding glass jolted me awake. The sound was so loud and so real that I assumed an actual car crash had just occurred right outside our house, or maybe across the river on McCarter Highway.

I climbed out of bed and hurried through the darkened house to the front door. My father was already there, holding the screen door open and staring out in the direction of the Passaic.

"You heard it too?" he asked.

I nodded.

We stood side-by-side in the doorway for a while, in the muggy darkness, scanning the silent river and its black bulkhead and the deserted highway on top. Nothing seemed amiss. There were no flames, no smoke,

---

6 Source: *Report to the Governor: Recommendations of the Passaic River Basin Flood Advisory Commission*, January 2011.

no sirens or flashing lights. Not a soul. Not a sound. Just the crickets. And the river, staring back at us.

We checked the newspaper the next morning and the morning after that. We watched the evening news. There were no stories about a crash. I guess it was a dream, a bad dream that we had dreamed together.

The 280-foot-long waterfall in Paterson is sandwiched between the Wayne Avenue Bridge to the south and the city's historic Hinchliffe Stadium[7] to the north. Topography divides the green sward surrounding the falls into separate spaces, like rooms. Atop the rugged U-shaped promontory to the north is the room that honors Mary Ellen Kramer. Overlook Park, where I met Ella, occupies the cradle of the J-shaped bluff that sits across the river and directly south of Mary Ellen Kramer Park. Further south and a few blocks east of the river is Upper Raceway Park, which honors John Colt's first raceway and the adjacent reservoir[8] of Passaic River water that surged down it on command.

From our starting point in Overlook Park, Ella led me on a short stroll upriver along McBride Avenue to a black steel footbridge that parallels the Passaic above the falls. The small Wayne Avenue Bridge was right behind us, a stone's throw upstream. A low, brick and wooden dam crossed the river just ahead. The dam held a calm dark reservoir of river behind it. There was a narrow outlet on the side of the dam nearest the footbridge. The outlet gate was open, and just a few feet below us the river

7 The 10,000-seat stadium was home to the New York Black Yankees of the Negro League from 1933 to 1945. Baseball Hall-of-Famers Josh Gibson, Judy Johnston, Oscar Charleson and James "Cool Papa" Bell all played at Hinchliffe, one of the few remaining venues from the old Negro League. Hinchliffe was also the site of performances by the Andrew Sisters, Duke Ellington and Abbott and Costello. Lou Costello was a Paterson native. The stadium was named for Paterson Mayor John V. Hinchliffe. Its Art Deco design was based on a plan submitted by the Olmstead Brothers of Central Park fame.

8 The reservoir was named for Stanley M. Levine, a Paterson native and member of the North Jersey District Water Supply Commission who died in 1989.

streamed through and fanned out across the wide, flat plate of basalt. We were standing at the very top of the falls. The Passaic gurgled and swirled for about 100 yards more before it disappeared over the sheer rock face.

The cliffs at Great Falls are made of trap rock, a kind of basalt that forms when super hot magma cools and hardens. The process can produce soaring columns of dark-colored rock. The New Jersey Palisades are trap rock. So is Devil's Tower[9] in Wyoming, the menacing mound scaled by Richard Dreyfuss and Melinda Dillon in the 1977 film *Close Encounters of the Third Kind.* "Trap rock" is an industrial term, not a geologic one. Crushed trap rock is used as the foundation for asphalt highways and parking lots. Trap rock can also be cut and polished for building veneers or headstones. Portions of the new Great Falls National Park are on the site of an old trap rock quarry.

I followed Ella across the black footbridge and onto a wide gravel path that continued north along the precipice. To our left, the Passaic flowed smoothly by at our level. To our right, 80 feet down, the river was aboil. The gravel path led to a second footbridge, the same green steel span that was visible from Overlook Park, the same bridge where my mother and I had our first glimpse of the falls.

The second footbridge connects the two basalt cliffs, spanning the chasm between them just below the cascade. (The French aerialist Philippe Petit, who got famous—and arrested—for illegally tightroping between the Twin Towers in 1974, high-wired across Great Falls near this very spot that same year.) The view was dizzying. We could stare straight

9 Although the Great Falls of New Jersey's Passaic River, the New Jersey Palisades and Devil's Tower in Wyoming share a similar origin (magma) and thus would all classify as trap rocks, the three formations have different modes of emplacement. Great Falls is part of the Watchung Mountain Range, which formed when molten magma erupted to the earth's surface in a series of lava flows, and then cooled. In the case of the Palisades and Devil's Tower, the magma stayed below ground, moving into adjacent sediments or rocks where it cooled and hardened beneath the earth's surface.

down into the trap rock abyss where the smooth black river had been churned into seething white foam. The canyon we were staring into cuts back for 280 feet, an incision in the rock. At the very back of the slash the two walls of the canyon almost touch.

When Ella and I crossed the footbridge to the U-shaped bluff on the other side of the incision, we entered Mary Ellen Kramer Park. I couldn't imagine a more dramatic vantage point, but Ella pressed on for another 100 yards to the very back of the canyon and one last overlook. The green space we were walking through wouldn't exist today if it hadn't been for Paterson's former First Lady. But the park that honors her was a disheveled disgrace. Ella, who worked with Kramer and is still working with Paterson officials to restore the area, couldn't contain her frustration. As we made our way along the path through the park Ella railed against the garbage and the poison ivy and the black wrought iron fence that she found so especially galling. "In 1972 when I came here with my kids it was a beautiful park," Ella told me. "But as soon as this *thing* [the fence] went up it was really the beginning of the neglect period."

The offending fence stood between the pathway and the falls like a row of black spears thrusting up through the earth. According to Ella, it was installed after a visitor stumbled on the pathway. Some kind of barrier is understandable since a misstep here could end badly. From the path, the grassy terrain slopes sharply down to the edge of the high cliff above the river. The iron spears have more aesthetic appeal than the razor wire-topped hurricane fences that defend the waterfront in Newark, but there's no question they mess with the view.

"We're not supposed to be in jail," sputtered Ella. "We're supposed to be looking at a beautiful natural landmark. Being able to commune with the falls is something that is unique in the world. But you can't see the falls without the fence. I've been crusading to get rid of this fence for the longest time." Unsuccessfully, it would seem. Ella's wrought iron enemy is slated for restoration not elimination in the current Great Falls rehab plan. The Passaic River Coalition is, however, working with the state to acquire two acres below the falls and eight acres above. If successful, the purchase will prevent a condominium complex from rising on the bluff.

Ella and I followed the iron fence back into the deepest, narrowest part of the canyon. Near the end we descended a set of steep stairs to a cramped landing that clung to the very edge of the cliff. Only the fence posts and 20 feet of misty air separated us from the mountain of water surging over the opposite face. We were eye-to-eye with the Great Falls of the Passaic River. Fence or no fence, the power on display was breathtaking and fierce at close range.

This fissure in the Watchungs at Great Falls is the natural dividing line between the upper and lower Passaic, between the cleaner, wilder upstream river and its more populated and polluted lower half. Prior to 1859, when the Dundee Dam was completed eight miles downstream in Garfield, Great Falls was considered the "head of navigation," the farthest point a ship— or a fish— could travel upriver from the mouth of the Passaic. At one time, before the pollution, before the Dundee Dam, the gorge below the falls teamed with indigenous fish. "There were sturgeon, shad, alewives, everything," said Nick Sunday. "You could catch fish with your bare hands." Not anymore.

My parents told us about my father's condition sometime during the fall of 1968. I was 16 years old.

"Your father has cancer" was the worst news we had ever been given. The moment those four words were spoken marked the abrupt end of what had been up to that point a really nice childhood.

My father had his first surgery that Thanksgiving. He came home from the hospital with a colostomy and a hopeful prognosis. A surgeon named Sarno removed most of his colon and routed what was left through a round walnut-sized opening in his abdominal wall. My father was a fastidious person. I remember thinking how hard it would be for him to adjust to this new reality.

But he did. He even developed a custom dressing for the colostomy, a small clear plastic baggy with a circle cut in the middle and a sterile gauze pad slipped inside. The hole was slightly larger than the colostomy. Once he taped the soil-proof dressing over the hole in his belly he was good to go. He enlisted us kids to mass-produce the dressings for him. We'd sit at the kitchen table like assembly line workers using a shot glass to trace a perfect hole on the baggy then cutting it out as carefully as we could with the pair of bandage scissors that my mother had saved from her nursing days.

It all just got to seem perfectly normal. My father went back to work. We kids went back to school. Life went on as before. The only real changes were that my father couldn't lift heavy things anymore, and although none of us ever mentioned it, we all carried around a worry that the cancer would return.

He was healthy for four years. In February 1973, he was back in St. Michael's Hospital in Newark. The surgery was short this time. The cancer had spread.

I visited my father in the hospital on an overcast winter day. He was lying in near silhouette in his hospital bed, as wan as the winter light. My brother Joseph, who was 16 by then, stood beside him at the head of the bed, still and silent. My father and my brother were holding hands.

My father came home a week after his surgery. My brothers helped him up the stairs to my parent's bedroom, where he remained until the day he died, in slow motion, four months later, in the early morning hours of June 29, 1973, in his own bed, with my mother at his side. For me, he died that morning in St. Michael's, holding my brother's hand.

We learned later that cancer was a Bruno family tradition. Intestinal varieties had claimed my father's father and two of his uncles, all before the age of 50. I blamed this uncanny susceptibility on a hold-it-all-inside nature—the Brunos didn't talk about their feelings—and a childhood spent in toxic Jersey City.

His death tore time in half for us. Forever after, our lives would be binary, sorted into events that occurred "before" or "after Daddy died."

Carl, Cathy and I won't be paddling close to Great Falls or its zigzag gorge today. Libby's Lunch is within sight of the falls but comfortably upstream. Edward Gertler says that it is "reasonably safe" during times of moderate flow— that is when much of the river's water is being hijacked by the hydroelectric station— to paddle almost as far as the Wayne Avenue Bridge before pulling out of the river above the falls. But this is not one of those moderate flow days. There is danger below the falls too. The river is so aerated by its plunge into the canyon that its surface becomes a froth, an illusion, like paddling on mist. On days like today when the river spills generously over the precipice, boats that venture too far up into the canyon can sink like stones.

We are done with our lunch, but we linger for a while. It's warm and cozy in Libby's and we're having a jolly time. I wish we could stay. But it's getting toward sunset and we have to get moving if we want to make the Fair Lawn take-out before dark.

We suit back up in the parking lot and trundle off single-file through the busy streets of Paterson, with our bright yellow life vests and our rolling boats. We may as well be from another planet. The site of us renders one teenager utterly speechless. She stands in the crosswalk, gaping and shaking her head slowly back and forth as she watches us cross the busy intersection at McBride Avenue and Spruce Street. We head northeast on McBride, past another gawking teenage girl. This one has her sullen younger brother in tow. She frets long and loudly over the wisdom of our kayaking adventure. "You be careful," she calls after us. "BE CAREFUL."

I feel like a celebrity, of the most minor variety mind you, but a celebrity nonetheless. Everyone we run into here in Paterson looks at us with such wonder and delight. It's like we're mascots, or maybe some kind of vision they're having. We seem to be awakening them to a possibility they had never really considered, a possibility that I had never really considered: that a person could actually take a boat out on the Passaic River. Stoked by all the love and attention and my charbroiled chicken sandwich, I bounce along behind Carl.

We follow McBride Avenue—a path that takes us between the river and the SUM's Middle Raceway—for about a third of a mile to Mill Street. The old Hamilton and Franklin mills face each other at the corner of Mill and McBride. The Essex Mill is one door north of the red brick Franklin building on Mill Street, right at the elbow where Mill bends east and becomes Van Houten Street. The SUM's Lower Raceway runs along Van Houten. Parts of its channel are clogged with weeds now. The lower raceway once sent Passaic River water coursing to Mill Street's Congdon, Phoenix, and Harmony and Industry Mills, as well as to the massive Allied Textile complex, which now lays in ruins at the end of the block.

Carl rejects Van Houten Street in favor of a shorter route back to the river. We cross Mill Street, pass the Salvation Army's Adult Rehabilitation Center and go left at Curtis Place. Curtis is a short angled street that runs due north and downhill towards the river. "Kayaks," mutters a shaggy-haired 20-something, nodding down at the boats as he saunters by us in search of an ATM. "Cool."

My enthusiasm begins to ebb as the sun remains a prisoner behind a wall of large clouds, the air grows cooler and our portage drags on and on, and then on some more. The put-in is still about a third of a mile upstream. Curtis Place curves east and turns into busy River Street, a four laner with cars whizzing past in both directions. We make a dash across River Street and follow Carl over the Mulberry Street footbridge to Alfano Island. The former SUM Island is a large, semi-industrial mound of dirt in the middle of the Passaic. It is home to the Alfano Furniture Company and, according to Carl, an illegal dumping ground for local landscapers. Large piles of wood chips and brush are lying around. The rotting-fruit aroma of fermenting mulch fills the air.

We maneuver the boats through a parking lot of 18-wheelers and down a brush-covered hill into a freshly-cut clearing. Two large Komatsu cranes, some earth-moving equipment and a few more big trucks are parked there in the soft dirt. Across the clearing, we can see an arched stone bridge with a footing on the island. This is the West Broadway Bridge, a graceful human-scale structure that is undergoing some kind of renovation or repair. Wood scaffolding encases the section of bridge that rests on the island. Construction workers have bulldozed a low earthen berm at the river's edge to keep the water out while they finish their work.

Carl points to the shallow backwater pool on the other side of the bridge, which is where we'll put back in. We roll the kayaks a short distance across the makeshift parking lot, over a narrow wooden plank that someone has laid across a large mud puddle and under the arch of the bridge. Carl has to duck to make the clearance. Once we're in the boats, we use our paddles to pole out to the deeper faster water.

Alfano Island is narrow here. It ends just ahead. The river's main channel races along on our right, about 30 yards away, confined behind the temporary levy. On our left and much closer is the quieter branch that loops around the back side of the island. We launch into this back-water flow, and with a few quick strokes merge quickly with the main channel at the island's southern tip. The Passaic is moving fast here. It carries us quickly downstream.

With a boost from the power of Great Falls, we cover the four miles to Fair Lawn in less than an hour, much faster than expected. We fly beneath bridges at Main, Arch, North and North 6th streets, and at Lincoln, Maple and Fair Lawn avenues and reach Memorial Park with daylight to spare. That last hour felt easy. We spent more time portaging today than paddling. But we overcome the temptation to paddle a little further, and instead load up the boats for the drive back to Camp Lane.

We leave Cathy in Fair Lawn. She'll spend the night at home in Bordentown, New Jersey, which is about an hour south. We'll meet up with her again tomorrow morning at the take-out point in Newark's Ironbound neighborhood. Carl and I retrieve my car from Camp Lane and make it back to our Totowa hotel by 7pm, just in time to see Magglio Ordoñez hit the three-run walk-off homer that beats Oakland and sends the Detroit Tigers into the '06 World Series.

9 | DUNDEE DAM

I AM BRACING FOR THE WORST on this clear, crisp October Sunday. The Upper Passaic was a delightful surprise with its riffled gorges, scenic waterfalls and peaceful loops through lush, empty bottomland. But the Lower Valley awaits us today, and it is not a particularly clean or scenic stretch of river, especially in the final 17 miles below the Dundee Dam. Paddling author Edward Gertler rates the so-called Lower 17 "fair to ugly" and "incredibly trashy." There will be sites to look forward to; the Nereid Boat Club in Rutherford is hosting its annual sculling races today. But I have lowered my aesthetic expectations as we prepare to face the grittiest, grimiest, most polluted part of the river. *My* Passaic.

As usual, today's journey begins where it will end, in Newark's Ironbound neighborhood. It's just after seven on Sunday morning, and the Ironbound's main streets are empty, save for the few well-dressed faithful making their way to 7:30 mass. We park two cars across from a soccer field in Riverbank Park. After we transfer all the gear, we pile into my rental and head north to Fair Lawn and Memorial Park, our put-in *du jour*.

It's a quick trip north up a nearly deserted Route 21. The highway claims all 14 miles of the Passaic's western shoreline between Newark and Clifton. We pass Riverfront Stadium, the $30 million home of Newark's minor league baseball franchise, which once threatened to replace Riverbank Park. We fly up the old McCarter speedway, past the Rutgers Street Bridge and along the eastern edge of Belleville into Nutley. We exit 21 at Route 3 East and cross the Passaic into Rutherford. We follow Riverside Drive (a.k.a. River Road) north through Rutherford, Wallington, Garfield and Elmwood Park to Fair Lawn. It's about 8:30am by the time we roll into Memorial Park.

There is a boat launch here. The wide dirt ramp looks a little neglected. Assorted trash is pressed inside several large muddy tire ruts. But it's an easy place to put the kayaks in the Passaic. We're on the river by 9, our earliest start ever.

The Passaic flows straight and south from Fair Lawn's Memorial Park. The river will execute a pronounced S-curve when it passes between the cross-river towns of Passaic and Wallington. (It makes a similar move further south when it hooks around the Ironbound.) Otherwise, we will head due south all day as we make our way through the Lower Valley and back to Riverside Park, about 20 miles downriver. The lone obstacle ahead is the Dundee Dam, which awaits us just four miles downstream in Garfield. Below the dam, the Passaic turns tidal. If we time it right, we can catch the outflow, which would make our last few hours of paddling a breeze.

The Passaic becomes a Great Separator here in its Lower Valley, forming a physical and socioeconomic divide between the larger, more diverse cities and towns along its western flank (Paterson, Passaic, Clifton, Nutley, Belleville and Newark) and the mostly smaller, whiter boroughs to the east (Fair Lawn, Elmwood Park, Garfield[1], Wallington, Rutherford, Lyndhurst, North Arlington, Kearny and Harrison). Close to a million

---

1 Although first incorporated as a borough in March 1898, Garfield, New Jersey became a city following a 1917 voter referendum.

people live along the river here. Some 550,000 occupy the west bank—more than half in Paterson and Newark—and 193,000 more live in the nine smaller communities along the river's eastern shore. With few exceptions, most notably the east bank towns of Rutherford and Fair Lawn, the Lower Valley cities and towns are not especially wealthy communities. For east bankers, the average median household income is $66,530 a year. That's 30 percent more than the average west bank household, but far less (by more than 50 percent) than the more affluent households upstream in Chatham or Basking Ridge.

Though east shore communities in the Lower Valley are becoming more racially and ethnically diverse— Asian and Hispanic minorities make up at least a third of the population in Garfield, Kearny and Harrison—they remain predominantly white. The west bank, by contrast, is more multicultural, owing to sizable African-American communities in Paterson and Newark, and to the mix of African- and Mexican-Americans in Passaic.[2]

The Lower Valley river towns are suburbs by definition; that is, residential areas on the outskirts of a city. In fact, they exist on the outskirts of three cities: Paterson to the north, Newark to the south, and New York City, which lies due east across the Hudson River. Compared to communities along the upper Passaic, the towns within this urban triangle feel compressed, compacted. There are more people per square mile, more homes and more cars. Houses are smaller, closer together, and nearer the street. There are parks in almost every town, some of them quite large. Three miles of Bergen County parkland wander the riverfront in Lyndhurst and Kearny. Still, Lower Valley river towns feel more and more urban to me. Residents seem to prefer pavement to gardens or grass these days. (The newer neighbors who moved in on either side of my childhood home

2 All population and demographic data for Lower Valley communities comes from the U.S. Census Bureau's 2005-2009 American Community Survey.

replaced their backyards with cement on one side and a built-in pool on the other.) Disease and development have felled many of the fine old oaks and maples that once lined the streets of my North Arlington neighborhood. In many cases the trees were not replaced, and the bare spaces left behind cry out for the stolen blessing of their shelter and their shade.

Although a bright October sun beams down on us, the morning air is brisk, and we take off paddling at a furious pace to warm up. The Passaic is uniformly wide here in the Lower Valley, about 200 feet across. I expect I'll miss the more intimate, sinuous stretches of the channel upstream. The Passaic promises a canal-like sameness here that reminds me of my years in Catholic school where conformity—behavioral and sartorial—was the rule. A thin strip of vegetation still insulates portions of riverbank in this upper part of the valley. Clearly visible behind the trees and shrubs is the dense residential/industrial footprint that bleeds out some 30 miles in all directions.

The landscape beneath the sprawl has an interesting geologic pedigree. The Lower Valley lies within the Piedmont region, one of New Jersey's four physiographic provinces. The Piedmont, or Piedmont Lowlands, reaches far beyond New Jersey. It extends from the state's northern border with New York all the way south to Alabama, varying in width from about 10 miles at its narrowest part around Newark to 125 miles near the border between Virginia and North Carolina.

The Piedmont is a transition zone between the rugged Appalachian Mountain chain to the west and the flat Coastal Plain to the east. Its placid surface, wide fertile valleys and soft rolling hills, made it an ideal place for farming and settlement. The gentle Piedmont was my home, and the adopted home of the Passaic as well. For while the river rises in the northwestern Highlands region, it flows through the Piedmont for most of its 86 miles.

Otto Zapecza is a groundwater geologist with the U.S. Geological Survey (USGS) in West Trenton, New Jersey. He manages the state's Geohydrologic Studies Program. At the time we spoke (in May 2006) his group was studying the effects of development on the Highland's aquifers.

Otto is a big guy. He stands well over six feet tall and has a bearish build. He grew up in the Piedmont too— in Garfield. As a kid, he loved fossils and dinosaur movies. He claims to have watched *King Kong*— the Fay Raye version— every time New York's WWOR broadcast the classic on its old *Million Dollar Movie* series, which was, like, a *lot*. Exotic rocks and fossils crowd the wide windowsills in his West Trenton office. There are trilobite burrows from the Delaware Water Gap, ropy black lava from Hawaii, a fine-grained "chert" from Britain's White Cliffs of Dover and a fossilized dinosaur footprint that Otto rescued many years ago from a Rutgers University gym where it was being used as a doorstop.

Strange as it seems, New Jersey is a great place for a rock lover like Otto. As you'll recall, the Garden State boasts not one, not two, but *four* different physiographic zones. It is rare to find this much variety in such a modest space. Neighboring Delaware, for example, is all Coastal Plain. "New Jersey has an incredibly diverse geology for such a small state," explained Otto. "The variety spans most of geologic time."

To describe the geologic forces that shaped the Piedmont province, and hence the Lower Valley, Otto began with the Valley and Ridge, the physiographic region that lies north and west of the Piedmont and defines the far northwest corner of New Jersey. The Valley and Ridge province is just that— an undulating landscape where high ground gives way to low in a regular repeating pattern, like terrestrial waves. The terrain in this part of New Jersey was once quite different. It used to be a broad flat plain of sand, mud and lime. These prehistoric sediments were deposited by the rivers and seas that bathed this part of New Jersey some 400 to 500 million years ago during the Paleozoic Era, or the "Age of Ancient Life," so-called because "life," in the form of primitive sea creatures and fish, first appeared in the Paleozoic.

The sediments laid down in the ancient Valley and Ridge province were compressed from above for many millions of years. The constant, crushing weight turned the sandy, muddy mash into rock solid layers of limestone, sandstone and shale. Then in the late Triassic— 230 million years ago, give or take— the continents began to move apart.

Continents are basically along for the ride. Like mountains and oceans and other landscape features that we can see, continents rest on top of massive moving plates, which are hidden deep beneath the planet's surface. These are the earth's tectonic plates, 70-mile thick slabs of crust, which bob like ice floes upon the planet's molten core. There are seven major tectonic plates. From largest to smallest they are: the Pacific, Eurasian, African, North American, Antarctic, South American and Indo-Australian. When these underlying plates shift position, the continents that ride on top of them shift too.

The late Paleozoic era, about 280 million years ago, was a restless time for plates and their continents. The North Atlantic and African plates were inching slowly towards each other. These two giants eventually collided, creating one huge land mass. Scientists call this ancient continent "Pangea," from the Greek meaning "all land."

The formation of Pangea was a slow violent event, a head-on, continental crash. The impact buckled the rock layers in New Jersey's Valley and Ridge province, rippling the once flat surface of the land into the accordion shape it bears to this day. (The resulting sine curve pattern is clearly visible in the road cuts along certain north Jersey highways.)

The province's erosion-resistant layers of sandstone would persist as ridges. The Kittatinny Mountain Range that rises between northwestern

New Jersey and eastern Pennsylvania is the most prominent example. The softer layers of limestone and shale would gradually surrender to the forces of wind and water and form the bedrock of the region's valleys. The Delaware River flows through a groove lined with limestone and shale.

In the era of Pangea, New Jersey sat cheek-to-cheek with Africa's Sahara Desert for millions of years, an adjacency both comical and fantastical and, in the realm of geologic time, ever so fleeting. The earth's tectonic plates are drifters. They continue drifting to this day. Pangea, the super continent, was destined not to last.

After we slip beneath the Morlot Avenue and Broadway bridges in Fair Lawn, Carl, Cathy and I come up to an old steel railroad bridge about two river miles downstream from Memorial Park. The bridge looks like a long battered boxcar up on blocks. There are two blocks on the west side of the river. The newer one is poured concrete, bland and putty-colored. Its partner, the piling closest to the water's edge—and, I suspect, the older of the two—is built of brownstones and is quite beautiful. The rocks are probably local. There were quarries just upriver in Paterson. The rough rectangular blocks are neatly stacked and mortared into a shapely, flat-topped pyramid. Pale green lichens mottle the sides of the dark stones. Two trees, leafless owing to the time of year but very much alive, rise straight and tall from the leveled top of the piling, having managed a toehold in whatever dusting of soil has collected there.

The railroad crossing is one of the 26 bridges that we'll paddle under on our way through the Lower Valley today. That's about a bridge per mile, roughly the same bridge density as the Upper Passaic. Some of the bridges are railroad bridges, like this one, that service NJ Transit and other lines. A few of the bridges have been decommissioned, like the old steel bascule bridge that stands in upright and locked position at the entrance to Newark.

Most of the bridges that cross the Passaic above the Dundee Dam, and all the bridges below it open in one way or another, a throwback to the time when the channel below Garfield was still navigable for large vessels. Mary Jane Brady, a childhood friend, lived on Fisher Place, a dead-end street that was a half a block closer to the river than our house. Her backyard ended at River Road. The Passaic was just across the street, on the other side of the high school football field. Mary Jane's mother would often stand at her kitchen counter looking out at the river from the window above the sink. She'd call for the kids to come watch whenever a big tug or barge lumbered upriver past the house.

Barges used to run fuel oil, construction equipment, sand and gravel and other cargo up and down the Passaic's lower reach. It's almost unheard of to see a big boat that far upstream today. But if the state succeeds in safely dredging dioxin-contaminated sediments out of the river in Newark, big boat traffic could return to this part of the river.

In Edward Gertler's view, the drawbridges are the only sites worth seeing in the Lower Valley. He admires their "control house architecture," by which he means the wheels and gears and joints that swing or lift or tilt the bridges open and closed. We get close-up views of these mechanics as we slip beneath the Lower Valley spans like grease monkeys rolling under car chassis. But what I like even better about the bridge undersides is the graffiti we find there— and best of all, the image of some guerilla artist, mindful of the tides, hunkered beneath these bridges like a talented troll, painting the dull concrete foundations with blue, red, yellow and sea green graphics. The work is invisible to the traffic that streams overhead. But from down here, at water level, it is a bold and welcome dash of color against the brown Passaic. More than that, the bridge murals are a sign of life in the Lower Valley and a rare touch of beauty on this "fair to ugly" stretch of river. They are works of art created for river people only, to be enjoyed by those who are daring or desperate enough to prowl the Passaic's "trashy" edge.

The birth of New Jersey's Piedmont region, like many geologic events, took place along an edge. Its gestation was long, slow and at times painful as the Piedmont's two parental plates pulled and slid and ground apart. But the outcome, the offspring of this wrenching tectonic farewell, was a verdant valley that would eventually become a haven for *homo sapiens*, a species that had yet to appear on the planet.

The edges of the earth's tectonic plates are dynamic and at times unstable places. The areas where different plates meet and sometimes touch are called boundary zones. They are the site of earthquakes and volcanoes and other seismic events that tend to alter or scar the surface of the land. Geologists characterize boundary zones based on the behavior of the neighboring plates. The boundary is considered convergent if the plates are coming together, divergent if they are moving apart, and transform if they are sliding past one another.

California's San Andreas Fault is a classic example of a transform boundary, which occurs when two plates—in California's case, it's the North American and Pacific plates—slide past each other in a horizontal direction. A convergent boundary can be found off the northwestern coast of the United States where the tiny plate of Juan de Fuca is not only inching towards its eastern neighbor, the larger North American plate, but actually diving underneath it. This "subduction" triggered the 1980 eruption of Mount St. Helens and is responsible for most of the volcanoes and earthquakes that characterize this seismically active part of the country, which I now call home. The divergent boundary is the kind we are most concerned with here, because New Jersey's Piedmont Province is the product of this type of tectonic motion.

The Piedmont began forming during the Triassic period, some 250 million years ago, when the once-conjoined African and North American plates began to move in opposite directions. The gap, or rift, between the two plates would eventually fill with seawater and become the Atlantic Ocean. Over eons, the ever-growing estrangement of the African and North American plates has transformed the Atlantic from a narrow channel into the vast ocean we know today. The Atlantic continues to widen as molten rock from the earth's core oozes up through the gap between the plates and adds more ocean floor.

But something else happened as the North American and African plates drifted apart. "There was a downfault," said Otto Zapecza. A slip. As the African plate moved east it slid down, "like an elevator," Otto explained, baring the adjacent face of the North American plate and leaving it higher and angled up in a westerly direction. This event produced the Piedmont, which we are paddling through today, a deep wide rift valley with a steep raked wall (the Appalachian Mountains) on its western flank. "The Piedmont is a rift basin," said Otto, a miniature version of the Great Rift Valley in Africa, which opened up some 35 million years ago when the African and Arabian plates began to yawn apart.

Once it formed, the Piedmont rift valley was shaped by powerful forces both above and below ground. The Triassic's torrential rains flushed sediment from the Highlands down the sluice-like western wall and into the basin. Lakes formed in the basin and disappeared, leaving behind their own rich organic deposits. As the continents continued to pull apart, cracks in the mobile landmasses allowed molten magma that was percolating through the bedrock to hiss to the surface. Some of that magma hardened into the three basalt ridges of the Watchung Mountain Range.

Despite its dramatic tectonic origins, the Piedmont rift basin evolved over time into a soothing landscape, a low-lying plain with soft hills and wide shallow valleys. The Piedmont became a safe, fertile, idyllic place which, eons hence, humankind would appropriate and utterly transform in the blink of a geologic eye.

This stretch of Passaic between Paterson and Newark is consistently wider than any portion of the river we've been down so far. And yet paddling here has me feeling a little claustrophobic. The riverbanks are bulkheaded for the most part and thick with homes, businesses, old mills and factories that loom above us and seem to press in. A few pocket parks provide some relief, and Carl offers a welcome distraction when, surrendering to his inner junker, he decides to salvage a large plastic kiddie car. He spends 10 minutes hoisting the once bright red car up out of the mucky shallows and heaving it on to the shore, insisting that he'll return to retrieve it.

Just as my claustrophobia begins to really gnaw, the Passaic widens appreciably and a long spit of land appears in the middle of the channel. What likely began as a shoal, or sand bar, has grown into a full-fledged island whose southern tip tapers into a long forested finger which points at the Route 46 Bridge, dead ahead.

Route 46 cuts across northern New Jersey, east-to-west, from the George Washington Bridge to the Delaware River. The road crosses the Passaic River four times along the way. It is the crazy windings of the river, and not the highway that create these multiple junctures. The span we're approaching is the last Route 46 bridge we'll paddle under on our river trip.

I can see the Garden State Parkway now. This is the only place where the Parkway crosses the Passaic, and beyond it the river widens yet again.

We have entered the lake behind the Dundee Dam. Dundee Lake is 300 yards wide and two-thirds of a mile long. The Parkway cuts the lake in half. Like low fences, the clear sharp lines of Routes 21 and 46 define the lake's western and northern borders, respectively. Until pollution rendered it unsavory in the early 20th Century, Dundee Lake was a favorite spot for swimming, fishing, boating and ice-skating.

A stiff headwind greets us as we paddle south into the lake. The wind blows unobstructed across the full fetch of this long, wide-open reach of river. Gulls wheel overhead. Rays from the bright midday sun ricochet off the surface chop, scattering sunlight everywhere and bleaching the entire scene. The sharp blue of the October sky, the sepia Passaic, even my day-glo orange Pungo fade in the gauzy glare that surrounds us like an aura.

We make our way over to a second elongate island on the western side of the channel. This one is marshy, ringed with dry brown grasses and topped by a low canopy of scrubby trees and shrubs that retain their summer green. The darker green exit signs along Route 21 peek out above the tree line. A large white refrigerator lies on its back, half buried in the shoreline sediment. Carl flushes a large blue ball from the marshy fringe. The ball's surface is covered with small studs. It looks like the business end of a Medieval mace. An impromptu round of kayak hockey ensues, the three of us slapping the ball back and forth with our paddles until Carl deflates our big blue puck and stashes it behind his seat. Another souvenir ball—that's six now, by my count.

The Dundee Dam can't be more than 300 yards downstream from here, but I don't detect the same kind of urgency to the current, the gathering momentum that seemed to draw us towards Little Falls and Great Falls. The lake must be disguising the speed of the river's current, its broad expanse acting like a sink to absorb and dampen the pace. The lake is doing a good job of it. I forget about the danger ahead until we come upon a barely submerged metal pipe about 100 yards upstream of the dam.

The pipe is about 10 inches in diameter. Like the dam up ahead, it runs across the entire river channel. It appears to be a safety barrier for the unwitting. A sign on the shoreline warns boaters not to pass this point, which would be nearly impossible anyway in a motorized craft. I suppose you could tilt your outboard up out of the water and let your boat drift over the pipe, but not many motorboats have a shallow enough draft for that. Even our kayaks get hung up. When Carl and Cathy paddle over the pipe their blue double catches and sticks and Carl has to use his paddle to work the boat free. My Pungo scrapes the pipe's curved surface, but it doesn't stop.

The pipe spooks Cathy. She asks Carl if we shouldn't take the boats out of the river here, now. Carl reassures her and Cathy, warily, paddles on to our planned take-out spot in Garfield, which is a mere 30 yards above the dam. There must be a healthy volume of water cascading over the dam today. I can hear the muffled roar as we pull the kayaks into shore.

We drag the boats up the low steep dirt bank, over a white guardrail and into yet another asphalt parking lot. This one, on River Road, used to belong to Garfield's Service Diner. Judging by the size of the parking lot, the Service Diner was a popular spot. It was torn down recently to make way for the new 10-acre Dundee Dam Park and Riverfront Greenway.

Plans for this $1.4 million project, a joint effort between Ella Filippone's Passaic River Coalition and the city of Garfield, include a park and a concrete boat ramp. The park is still very much a work in progress. There's no greenery in sight, just a large yellow crane sitting idle on the torn up tract north of the parking lot where used car lots and an auto graveyard once stood. It's unsightly now. But when the Greenway project is complete, residents of the garden apartments here on the corner of

Division Street and River Road will have an unobstructed and much more verdant view of the Passaic.

There's no sidewalk on this side of River Road, and barely any shoulder. So we roll the boats across the street, past the dark-haired man standing cross-armed on the front lawn of the garden apartments, and leave them on the sidewalk in front of Riverview Auto Sales while we reconnoiter the put-in below the dam.

According to the city of Garfield, the Dundee Dam was built on the site of a seven-foot-high natural waterfall. The first dam was completed in 1838 and replaced by the current stone structure in 1859, which makes the Dundee Dam older than Garfield itself. The dam is jointly owned and operated by United Water, a Bergen County water supplier, and the North Jersey District Water Supply Commission. Its terraced face, which was recently renovated and reinforced, is 20 feet high and 450 feet wide. It spans the Passaic between Garfield and Clifton.

Praised as an engineering feat in its day, the Dundee Dam was supposed to make the Passaic more navigable. The dam's designers installed a large gate and a two-mile-long canal on the Clifton side of the river. The plan was to divert river water and ships into the canal, which followed the river's western shoreline north through Clifton and Passaic all the way to Paterson. As it turned out, the Dundee Canal owners went bankrupt two years after work was completed, and the canal was barely used. But Dundee Dam did come in handy for powering Garfield's growing textile and paper industries, which in the heyday included Garfield Worsted Mills and the New Jersey Spinning Company. Dundee Lake remains the sole source of water for the Marcal Paper plant in neighboring Elmwood Park. Marcal, the toilet paper manufacturer, is one of the last industries still operating along this stretch of Passaic.

Garfield is geologist Otto Zapezca's hometown. Otto and his childhood buddies spent a lot of time along the Passaic here. "There were tons of places we would frequent on River Road," said Otto, ticking off the Service Diner, the Sub Base (an old sandwich shop), the bowling alley and Colman's Golf Land, a compact 18-hole golf course with a driving range whose soaring nets kept the golf balls out of the Passaic. Sometimes Otto and his pals would hang out by the old dirt pullout below the dam where locals would scoop up buckets of river water and wash their cars. "We did a lot of fishing too," said Otto. "If you crossed the Passaic at the Outwater Lane Bridge, there was the old canal and canal path on the other side, and we'd go back in there fishing for carp and snapping turtles. The carp were huge. We'd bait these giant gang hooks with dough and throw them in. I worked in a bakery at the time so I had an endless supply of dough."

Most of Otto's old haunts are gone now. But when Carl and Cathy and I cross River Road to get our first close-up look at Dundee Dam, we are delighted to find that, thanks to the Greenway Project, the old dirt pullout/car wash has not only survived but has been upgraded to a gleaming white cement boat ramp that rolls ever so gently down from River Road to the foot of the dam. It is the Rolls Royce of put-ins.

There's just one hitch: the entrance to the ramp is cordoned off by orange mesh fencing, the kind you often see at construction sites. We check around the area for "Do Not Enter" or No Trespassing" signs, but we don't find any. Nor do we see any obvious hazards. The ramp is clearly finished. It's practically begging us to inaugurate it. After a very brief confab we rationalize: the construction crew probably just neglected to remove the fence.

We roll the kayaks back across River Road. As we finish handing the second boat over the orange mesh fence a Garfield police cruiser pulls up beside us. Cathy and I exchange a knowing glance. The guy on the lawn of the garden apartments, the stocky, dark-haired, middle-aged, scowling snoop, must have called the cops. We're busted.

I prepare to explain my book project and plead our case to the young officer who steps out of the black-and-white. He seems like a reasonable fellow, though his eyes are hidden behind a pair of dark aviator shades. He asks us what we are doing. He makes a call on his cell phone. He confers with his partner, the driver of the car. Finally, he waves us on. With great relief, and our own Garfield PD escort, we relaunch the Pungo and the Perception into the whitewater below the Dundee Dam. I go first. My bow bucks over the colliding waves as I head for the middle of the channel. Whenever the boat noses a trough, the crest breaks over us and we take on a little Passaic.

10 | THE NEREID

OUR LAST PORTAGE IS BEHIND US. There's nothing but river now between here and our final take-out in Newark. As we leave Garfield we pass the mouth of the Saddle River, the Passaic's last major tributary. It enters the mainstream just south of Garfield's Passaic Street Bridge. Beyond the Saddle River the Passaic takes an S-curve, snaking north, then south again between the communities of Wallington and Passaic.

Just north of Wallington's Gregory Avenue Bridge we pass the place where George Washington and the remnants of his Continental Army beat a hasty retreat across the Passaic in November 1776. Washington was fleeing the British, who had already chased the colonials out of Brooklyn into Harlem, and then across the Hudson River to Fort Lee, New Jersey. What was left of Washington's army abandoned Fort Lee on November 16th. They left with little more than their weapons and the clothes on their backs. Many soldiers were barefoot.

Washington was headed for Acquackanonk Landing, the major trading port that thrived, during the colonial era, along the eastern edge of the city of Passaic. Washington's army crossed the Acquackanonk Bridge with the British in hot pursuit. The team of local men and boys, most of them farmers, who quickly dismantled the bridge, probably saved the Revolution.

With the bridge gone, the pursuing British troops were forced to ford the Passaic upstream in Garfield. Heavy rains and high water delayed their crossing for a week, allowing Washington time to push into Pennsylvania where, a month later, he would stage his daring Christmas Day crossing of the ice-strewn Delaware River. That surprise raid against the British and Hessian troops encamped on the Jersey side of the Delaware turned the tide of the Revolutionary War.

The markers that commemorate the events surrounding Washington's historic Passaic River crossing are modest, to say the least. Both are located upstream on the Garfield side. They are a tall stone monument—it looks like an outdoor fireplace—in a little park near the river at the foot of Columbus Avenue, and a small plaque on the side of the red brick YMCA building at Outwater Lane[1]. There is nothing at the actual crossing site, save for a few surviving patches of green on the Passaic side of the river, shoehorned into the knot of highways and industrial parks that now chokes the bank where Washington billeted his troops on that cold, wet November night.

We continue south, always south now, paddling under Wallington's Main and Union Avenue bridges and down into the Rutherford-Nutley reach, the stretch of Passaic that claims the poet/pediatrician William Carlos Williams (born and raised in Rutherford) and media superstar/jailbird Martha Stewart (a Nutley native). North Arlington, which as far

---

1 The inscription on the stone monument identifies the place along the river as "Post Ford," a site "Frequently used by both armies during the Revolutionary War." Adrian Post was the Colonial-era owner of the farm and gristmill on the property. The YMCA plaque commemorates "Washington's march in 1776 along a route bordering lands now forming the City of Garfield."

as I know has never spawned a celebrity of any stature, is still about five miles downstream.

Near the southern edge of Rutherford, just a half mile from the Lyndhurst border, we come to the Nereid Boat Club. Suddenly, we are not the only boats, or the only people, on the river.

A crowd of 200 or so mingles on the grassy bank outside the wooden boathouse. Single scullers and teams of two, four and eight hoist their needle thin racing shells overhead and march them down the steep wooden ramp to the club's long floating dock. More spectators look on from the boathouse's bay of second-story windows. They're all here for the Head of the Passaic Regatta. The Nereid's annual autumn weekend competition draws rowers and fans from all over the Northeast.

Growing up in North Arlington, we'd occasionally see high school crew teams from Kearny, Nutley or Belleville out training on the river. They still do. But the Head of the Passaic Regatta is a recent sculling *event*. The large gravel parking lot beside the gray and white boathouse is cluttered with racks of racing shells and small white tents where vendors hawk hotdogs or t-shirts or river causes. When our kayaks inadvertently drift onto the course, two volunteers in an Army green skiff motion for us to move to the opposite side of the channel. We paddle quickly aside as a brace of four-man sculls glides towards us up the Passaic.

The 132-year-old Nereid[2] Boat Club is the last of the original Passaic River rowing associations, the sole survivor of New Jersey's once proud sculling tradition. Until the 1940s, rowing was a hot sport along the Passaic. Competitions like the Head of the Passaic spectacle that Cathy, Carl and I are witnessing today were regular weekend happenings. Ralph Van Duyne, Chief Engineer for the Passaic Valley Sewerage Commissioners, recalled the heyday of Passaic River sculling in a 1946 interview with the *Newark*

---

2 In Greek mythology, Nereids were water nymphs that often accompanied Poseidon.

News. "Sixty years ago I used to go in swimming near the old Triton Boat Club at the foot of Chester Street [in Newark]," Van Duyne told the *News*. "At that time, the Passaic was famous for its Middle States Regatta, with thousands of people lining the banks to see the competition."

Besides the Triton rowers, Newark also claimed the Institutes, the Mystics, the Eurekas and the Passaics. Even the smallest river communities had their own rowing clubs and the competition was football fierce. Local residents took to the banks on summer weekends to cheer their hometown scullers. Pollution and a 1940s-era race-fixing scandal saw the decline of the sport and the disappearance of most rowing clubs. Passaic River pollution got so bad, said Van Duyne, "that freshly painted boats would tarnish overnight. …[F]actories along the river had to close up for weeks during the hot weather. Their employees couldn't stand the odor." But the Nereid Boat Club survived.

When I began calling various club officials to find out exactly how the Nereid had managed to avoid extinction, everyone told me to talk to Erik King. I caught up with him in October 2005, right after he and his partner had won the mixed doubles competition in the Nereid's third annual Head of the Passaic Regatta. Erik and I sat on makeshift wooden bleachers amidst the crowd of cheering fans that swarmed the riverbank on that summery October afternoon.

Erik looked like a rower, broad and strong across the shoulders and chest. His hair was clipped short and turning prematurely gray, which made his blue eyes seem even bluer. Despite a busy schedule—Erik is married with two children and a successful contracting business—he rows every day, year 'round, even, rumor has it, when there's ice on the Passaic. "It's a great escape," he said, about his obsession. "When you're rowing you have to concentrate so deeply that you really can't be thinking about anything else. You're rowing along this beautiful river. You have this beautiful feeling. I've had days where I had a row that was so great that if I was in a bad mood it totally turned me around."

Erik King has a long and deep connection to the sport and to the Passaic. As a teenager, he competed for Belleville High School. His father rowed on the Nereid teams of the 1940s and '50s, back when the Nereid clubhouse was located farther south, near the Belleville-Newark border. That boathouse was eventually floated upriver, by barge, to a small waterside park in Belleville. It was used by the Nereids and by both Belleville and Nutley High School crew teams until an arson fire destroyed it in 1962. Route 21 runs over the site now.

The first Nereid Boat Club formed in Newark on May 15, 1866 and disbanded several years later. But the current Nereid Boat Club traces its origins to the summer of 1875 when eight Belleville rowers—John C. Lloyd, Charles Leverich Webster, William H. Webster, James D. Ferris, Clarence S. Van Houten, J. Roger Kingsland, Joseph Kingsland and William M. McCreery—resurrected the Nereid name. The club went "dormant" again after the 1962 boathouse fire, said Erik, but it didn't die. The dream and all the paperwork lived on with Homer Zinc.

Zinc was a lawyer, a rower, a Belleville native and one of the Nereid club's officers. (He was also a classmate of my mother's at Belleville High.) He eventually left Belleville for Princeton, New Jersey, but he kept the club's charter. In 1988, determined to revive the Nereid rather then start a new rowing club, Erik King traveled to Princeton to try to convince Homer Zinc to transfer the club's original charter. "I met him at his house," recalled Erik. "He was older and he wasn't healthy, and he didn't really trust handing anything over to me until he saw that this was legitimately going to happen. So he just kind of brushed me off."

The brush-off was understandable given that Erik could only produce 12 committed club members, including himself and his wife. But during the next year, Erik worked his old Belleville High crew connections and the membership grew. When it hit 100, he went back down to Princeton. Zinc had suffered a heart attack in the interim. The two met in his hospital room this time. "When I showed him what I had done and how serious we were," said Erik, "and with him knowing it was now or never, he signed everything over to me."

With Zinc's blessing and the original charter the neo-Nereids moved into the Rutherford boathouse several years later. The old two-story building once housed the Rutherford Yacht Club and after that a Chris Craft dealership. It had been empty for some time when the Nereid club members invested their own money and sweat equity in renovating it. The club now offers rowing classes and programs, kitchen and shower facilities, equipment and storage, a big stone fireplace, a river web cam and, of course, its annual autumn head racing festival.

In "head" racing, rowers compete against the clock. The boat with the fastest time in any particular heat wins. This style of racing is much more practical on the Passaic, which isn't wide enough to accommodate more than four shells abreast. Head of the Passaic competitors leave the Nereid's two floating docks and row downstream to the starting line at the Stack Restaurant in North Arlington. (I've enjoyed many a charbroiled cheeseburger at The Stack over the years.) The boats queue up below a series of yellow buoy pairs that define what's called the "chute," an alley about 30 yards long that acts as a kind of warm up lane. Two red buoys at the upstream end of the chute mark the starting line. Volunteers in boats and on the shoreline use bullhorns to tell each boat when it has entered the chute, and then when it has crossed onto the actual course. The race ends three-and-a-half miles upstream at the Nereid boathouse.

On that Indian summer afternoon in October '05, I watched the fourth and last block of racers from the Michael J. Supple Park, a remnant of green just south of the Route 3 Bridge in Rutherford, about a half-mile from the finish. The low, split rail fence that separates the park from River Road was perfect for perching. A tall beech tree provided shade and cover. Its drooping branches framed the river and the rowers as they powered past me under the Route 3 overpass. It was late in the day. The tide was out. The river was low. The wide exposed bank was slathered in the dark pudding mud I remembered. The sculls hugged the near shore where the channel was deeper. Despite the traffic coursing along McCarter Highway on the opposite shore, the park was peaceful. It wasn't hard to imagine sitting in this very spot on a race day long ago, picnicking with fellow spectators, waiting for Homer Zinc and his Nereid crewmates to glide into view.

Erik King was adamant: the only way to get a real feel for what it's like to row on the river is to row on the river. So on a warm May morning the following spring, he took me sculling on the Passaic. The air was still that morning, like the surface of the water. I had never been in a racing shell before, but Erik was a thorough, patient teacher. He showed me how to secure the green and white oars, or "blades," into their riggers. At his direction I stepped into the white racing shell at the only proper place—the small blue rectangle painted in the very center of the boat's bottom—and lowered myself into the sliding front seat of our double. I Velcro-ed my feet into the white sneakers that were themselves secured to two angled metal plates in front of me. That gave me a stable push-off point. It also attached me firmly to the boat. Before I could contemplate too deeply the implications of having my feet strapped to a tippy scull,

Erik hopped in behind me and pushed us off. After that, true to his word, I did all the work.

Rowing the scull required slightly more technique than rowing the old Andover skiff. For the "drive phase" of each stroke Erik had me stretch forward until I was doubled in half, then gently dip my blades into the river behind me and explode backwards, pushing off with my legs and pulling the blades through the water in one clean shallow sweep. For the "recovery phase," I had to bring the blades up out of the water and move them back to cocked position. That was surprisingly hard. The wooden oars in our Andover rowboat were short. The scull blades were really long. At the end of every drive, the handles of both blades would collide in the vicinity of my lap. By the end of our hour-long session my thumbs and the back of both hands were bruised from all the banging.

The biggest challenge was just to coordinate all the little details of technique. Like twisting my wrist slightly as I brought the blades in and out of the water so they'd be positioned properly, either perpendicular to the water's surface for the drive, or parallel for the recovery. There were moments, six or maybe eight consecutive strokes, when I managed to put it all together and the effect, the speed, was exhilarating. For 30 seconds or so we defied the laws of friction and skated across the Passaic's placid surface. Unlike kayaking, where you actually sit below the water line, sculling has you sitting at the surface. It was like riding a cushion of air. I didn't have to pull hard to generate speed. It was more about rhythm. *Reach, dip, pull.* Legs, arms, back, wrists all working together. Team Mary. It was a rush. I could see why Erik is addicted to the sport.

But why the Passaic, I asked him? Wouldn't the rowing be better on a cleaner patch of water? Didn't he worry about, you know, dioxin?

"The river can be scary, and a lot of people are scared," said Erik, who seemed nonetheless sanguine about the dioxin. "It's hard to say what the worst thing I've seen is." Though the floating bag with the dead animals inside was right up there. So was the day Erik came out to row and there was so much debris on the river that he couldn't even get his boat into the water. "It was solid garbage," he recalled. "You could walk across."

The Nereid was actually forced to cancel its 2007 regatta because the river was so inundated with junk. The Passaic Valley Sewerage Commissioners operates two vessels whose job is to skim garbage off the river's surface. But even the skimmer boats couldn't save the '07 race.

Bob De Vita ran the skimmer boat operation. De Vita, who died in September 2009, worked for the PVSC for more than 30 years, and managed the skimmer boat program since it began in 1998. In an average year, De Vita told me, the skimmer removes about 200 tons of "floatables" from the river. One year, the take topped 280 tons. "A lot of the weight is with the timber," said De Vita. Riverside trees, chain-sawed logs, telephone poles and driftwood from old piers and railroad ties are common collectibles. The biggest bobber his crews ever hauled in was a dilapidated barge that broke loose from its moorings in 2008. "This thing was monstrous," said De Vita. "It was like eight tons. It had been out there for so long that the wood rotted and trees started growing into this thing. It looked like a floating island." Skimmer crews encounter dead bodies from time to time. "One guy had a bullet in his head," said De Vita. "We don't pick them up. We just call the cops." But in terms of sheer volume, the biggest haul by far is plastics—plastic bottles, plastic take-out containers, plastic packaging. Plastic, plastic everywhere. "I stopped drinking my little Poland Spring," said De Vita. "I don't use plastic bottles anymore."

When I asked Bob De Vita what went wrong for the Nereid's '07 regatta, he pointed to an unfortunate confluence of events. First there was the weather. Storms help to flush debris down the Passaic, but the autumn of 2007 was bone dry. "From mid-August until October it never rained," he recalled. "I've never seen the river so covered with stuff." Even so, the skimmer boats probably could have handled the excess, *if* they had been able to reach the Nereid. Unfortunately, both boats were effectively out of service that weekend.

The little skimmer was at the bottom of Newark's Weequahic Lake, the victim of a pontoon puncture it suffered during an operation to clear the lake of aquatic plants. The big skimmer was operational, but to travel upriver from its berth in Newark to the Nereid the big boat needs the drawbridges along the way to be operational too. "To get to Rutherford I need to open four bridges, " said De Vita. "If one's down I'm out of business." There were bridge problems that weekend. "I couldn't get the big boat up there."

Dry weather, broken bridges and one pierced pontoon all had a part in canceling the '07 regatta. But the real culprits were the hundreds, or thousands, or maybe even millions of people living along or near the Passaic River who just can't seem to secure their garbage. In fact, said Bob De Vita, the one simple thing that every one of them, that every one of *us*, could do to affect an immediate clean up of the Passaic River is "not litter."

The intermittent trash does not deter Erik King. He remains devoted to the Passaic, insisting that the river has much to recommend it as a venue for rowing. The Passaic, for example, "is not too big, which keeps it from getting too wavy so you can pretty much row every day," said Erik. "It's brackish, which is good because real salty water is bad for the boat parts. It's good that the river is tidal and moves in and out, because that keeps it from getting stagnant so it doesn't freeze early in the year and we get to row on it a lot longer."

I would add one more plus to Erik's list: the Passaic is pretty. I can't believe I'm saying this, especially about the Lower Valley reach of river. But beneath all the metal bulkheads and plastic trash and rank mud, behind all the heinous abuse and degradation, there remains something utterly lovely about the Passaic. I noticed it during my first trip up the river on the Aqua Patio with Andy Willner and Bill Sheehan. I see it now in the sensual curve of the channel, in the sheltering embrace of the riverside trees, in the sigh of the tide and the glint of the light and in the river's constant quiet yearning for the sea. If I listen carefully I can almost hear the river breathing. It is alive. Still. Always.

Paddling downstream from the Nereid, my kayaking companions and I pass scull after scull of competitors. They power past us with their long elegant blades swinging fore and aft in pinpoint synchrony. We, by contrast, labor against a still incoming tide. I can feel its drag on my broad bus of a boat. Whenever I stop paddling I start to drift back upriver. We're still a good five miles from Newark. With resistance like this, it might be tough to make our take-out by nightfall.

Before I can share my concerns with Carl, I'm distracted by the smell of barbecue. We are crossing into Lyndhurst now. There's some kind of fair going on in the County park along the river, and it smells too good to pass up. We pull the kayaks to shore just downstream of the large rock—two slabs of concrete actually—where the teenage Ella Filippone chatted away her Sunday afternoons.

We head straight for the food. Cathy treats us all to grilled pork teriyaki sandwiches on pita bread. There's a live band whose front man looks like Steve Van Zandt. While we eat, we listen to him croon *Hotel California* and chat with a middle-aged woman from Union who's selling dried flower arrangements with a Yuletide theme. She thinks the fair organizers made a mistake by not setting up on the opposite, River Road side of the park where they would have attracted more drive-by traffic. True enough, we tell her, but then you wouldn't have gotten the kayak crowd.

As we walk back to the boats, I talk tide with Carl. He registers my concerns, and my mutinous suggestion that we could, if worse comes to worst, take out in Kearny, at the high school boathouse near the Rutgers Street Bridge, and cab it down to Newark to get the cars. But he remains unmoved.

We set off once again on the river. Within the hour, the resistance suddenly vanishes. The paddling gets easier. Carl is right. The tide has turned. We ride the outflow into North Arlington.

11 | BATHURST AVENUE

THE PASSAIC HAS NEARLY RUN ITS COURSE by the time it forms the western boundary of my hometown. The Borough of North Arlington is the southernmost town in Bergen County, the New Jersey County that occupies the northeastern corner of the state. North Arlington, about two-and-a-half miles square, is just five miles north of Newark and, depending on traffic, about 15 minutes west of Manhattan. A third of North Arlington is taken up by the Holy Cross Cemetery, a shady tract on the ridge overlooking New York City where my father and mother are buried and where I once spotted a Canada warbler. A quarter of a million people are interred in Holy Cross cemetery, which means that in North Arlington today the dead outnumber the living by about 16 to 1. The rest of the town is a jigsaw of modest single family homes and garden apartments, four elementary schools, a middle school, two high schools (one public, one Catholic), five churches, the Little League baseball field, high school baseball and football fields, a public library and four large chain drugstores, a retail testament to North Arlington's slowly graying demographic.

I'm not sure why my parents chose to settle in North Arlington. I suspect it had something to do with location—the town was close to both their families, but not too close. North Arlington was also an aspiring place, lots of young families, pretty good schools, a small town atmosphere in the shadow of New York City. And the housing must have been affordable because after just two years of apartment living—in an upstairs flat on Webster Street—my parents bought their first and only home at 118 Bathurst Avenue.

It was a short street, just three blocks long, which paralleled the river. One block below Bathurst, at the very bottom of the valley, was River Road, a two-lane blacktop that followed the gentle curves of the Passaic. Small businesses—restaurants, gas stations, the North Arlington Diner, Charlie's Nursery & Garden Center, Homelite—shared the water side of River Road with a few public parks, and the town's ball fields. The other side of River Road was almost entirely residential.

One half mile or so above Bathurst Avenue, at the very top of the hill, was Ridge Road. This was North Arlington's main drag, named for the ridgeline it followed. I spent many a high school night driving up and down Ridge Road in a blue Cutlass convertible that belonged to my friend Terry's father, the cigar chewing Mickey Filippone (no relation to Ella). He owned Melrays, the large furniture emporium on Ridge Road. We called Terry "the furniture heiress."

From the ridge top in North Arlington, you could look east across the Meadowlands, famed graveyard of mob victims and the fine marble remains of New York's old Pennsylvania Station, and admire the Manhattan skyline, either glistening or twinkling depending on the time of day. If you drew a straight line from my childhood home due east, it would cross the Meadowlands and connect with Manhattan Island at West 82nd Street, about eight miles as the crow flies. I remember standing on the ridge once and taking in Manhattan. It was a clear morning in December. I was 24 years old, a graduate student in aquatic ecology, home for the holidays. That was the first time I realized that the Meadowlands was a salt marsh, and not just a convenient place to get rid of garbage and the occasional snitch.

The Bathurst Avenue of my childhood was all single-family houses in a mish-mash of architectural styles. The tidy blue Cape Cod on the corner belonged to the Harz family. Mr. and Mrs. Bryan, retirees with grown children, owned the dignified ranch style house at the other end of the block. It was red brick with leaded-glass casement windows, and a small detached greenhouse in the back. In between, and right in the middle of the block was our own faux Tudor, its twin peaks encased in pearl gray stucco.

Bathurst's architectural treasure, directly across the street from our house, belonged to Alice Parker. It was a two-and-a-half-story, 19-room Victorian mansion, and though it towered over the rest of the neighborhood, the Parker house never felt haughty or out of place. It had a handsome mansard roof and a wide wraparound porch on the ground floor that was accessed through several sets of tall, white French doors. There was a grand central hall with a winding staircase and stained glass windows. There were black walnut doors throughout and four Italian marble fireplaces, and all the many bedrooms on the upper floors had balconies with views of the Passaic. In one of our many acts of trespass, we discovered an abandoned concrete fishpond behind the house. The lawn that sloped gently down to the pond had a lone mimosa tree placed just so at the corner farthest from the street. We would have gladly forsaken the Homelite field to play our football on that lawn. But the ever-vigilant Alice Parker wouldn't let us anywhere near it.

Years later, in the early 1980s, after Alice died and her only son Tommy sold the house to a local developer, I found myself at a town council meeting with my old Bathurst Avenue neighbors. It was a show of solidarity to try to block the developer's plan to raze the Parker place and put three, multiple-family houses on the property. Most of the neighborhood turned out, even prodigals like me who hadn't lived on the block in years. My youngest brother Paul came with me. The Bruno delegation. My mother still lived on Bathurst at the time, and would for another 20 years.

North Arlington didn't seem to have any zoning laws back then, so our strategy that night was to sway council members with arguments about protecting property values and preserving the character of the neighborhood and if that failed to raise the your-block-could-be-next specter.

Midway through the open mic portion of the meeting a young man from the North Arlington Historical Society—indeed, he was the North Arlington Historical Society—rose to speak. His name was Robert MacFadyen. It turned out he had known my father. They had worked together on a book commemorating North Arlington's 75th anniversary.

At the meeting that night, Robert MacFadyen surprised the council members, and the rest of us, by revealing that Alice Parker's house was the last of the Passaic River mansions. It was built by James B. Hervey, a retired merchant sea captain, in 1873, three years before General George Armstrong Custer made his ill-advised last stand at Little Big Horn. According to MacFadyen, West Point cadets were among the many guests who attended grand parties at the Hervey mansion. So my father had been right after all: Bathurst Avenue was coveted water view property, a real high society haunt. He and my mother had just been 100 years too late.

Alice's father, Joseph Parker, purchased the mansion directly from the Herveys in 1909. "The Parkers knew the Herveys," said MacFadyen when he and I met for lunch in 2006. "They even told about how the Hervey boys, when they went wenching, would crawl out and shimmy down from their bedrooms and swim across the river to enjoy the pleasures of the wild women of Belleville."

MacFadyen was working for the Bergen County Police Department back then, in the records division. He has since retired. He had lost his sight as a result of diabetes. Like most of the historians I spoke with, he held a dim view of the Garden State's commitment to historic preservation. "We've trashed all our history in New Jersey," he said, as he munched a hamburger with onions at a diner near the police station. MacFadyen's mood brightened, though, when he recounted life along the 19th Century Passaic.

From his mansion, said MacFadyen, Captain Hervey would have been able to survey his holdings on the river's eastern shore and enjoy sweeping views west across the Passaic. Hervey's property extended uphill from the river all the way to what is now Ridge Road. The view encompassed his carriage house and barns, his icehouse and springhouse and to the southeast his Pacific Steam Laundry[1] where Chinese workers cleaned and pressed men's shirts for shirt manufacturers in New York. Hervey may have glanced down at the yachts he kept moored near the Rutgers Street Bridge, or counted the growing number of mansions across the river in the bustling village of Belleville. He might have admired the drape of the willow trees along Belleville's streets, or the elegant cut of the wooden sloops loading and unloading at the village docks. Many of the boats would have been built just upriver at the Jerolemon Shipyard in Lyndhurst. Today, the Parker mansion on Bathurst Avenue is all that remains of this scene and this era in the life of the Passaic.

---

1 The steam laundry that Captain James Hervey built on Stevens Place in North Arlington in 1867 survived well into the 20th Century. It was a three-story red brick building with large windows and slate sills. As kids, we used to walk past it in the summertime on our way to Stevens playground, a former graveyard, which was right next door to the old Hervey laundry. By then (the early 1960's), the plant had been taken over by the Diamond Boning Corporation, one of only two companies in the U.S. that made steel supports for surgical corsets. The Diamond factory always reeked of solvent. Despite its historic significance, the old Pacific Steam Laundry, along with Stevens playground, was demolished in 1987 to make way for a small residential development.

On the night of North Arlington's council meeting, the night that decided the fate of the last Passaic River mansion, the young Robert MacFadyen concluded his remarks by imploring the council members to consider the property's historic significance before granting the developer's request. His plea was eloquent, and perhaps because of it the Parker house was spared. But its grandeur and grace were much diminished by the compromise that the town council struck with the developer. Two brick houses, boxes really, surrounded by cement, now stand where Alice's lovely green lawn and mimosa tree used to be. To make way for the boxes, the builders also tore out two old maples and one of the two locust trees that used to shade that side of the street. In winter, when those trees were bare, we too could see the Passaic from our front porch— it was there, across Alice's lawn, past the mimosa tree, down the hill, across River Road, just the other side of the high school football field.

From my kayak, I scan the hillside for the Parker house. The ball fields we are passing by are directly below the old mansion, but Alice Parker's place remains hidden from view above me, behind the trees. I get nostalgic at the sight of the Little League field and the adjacent high school football and baseball complex. They were the scenes of many childhood triumphs. My brother Johnny's outstretched touchdown grab in a high school football game. The no-hitter Joe pitched in a Little League contest. Paul's three-hitter to capture the North Arlington Little League World Series title. My grand slam to beat a rival in the town's Girls' Softball League. Exquisite memories, as clear and sharp in our minds as if they had happened yesterday. They happened long ago, before my father died, before we tasted sorrow, before we lost our luck.

In truth, my hometown doesn't feel like home anymore. How could it without my mother, who died in 2003. We sold the family house a few months after her death. The man who bought it fixed it up and flipped it within a year. I don't know who lives there now. I haven't been by 118 Bathurst Avenue in a long time.

None of us lives in North Arlington anymore. Johnny is closest, across the river in Belleville. My sister is in Pennsylvania. Paul and Joe are 20 miles south in Cranford and Westfield, New Jersey. I am alone on the other side of the country.

Our physical separation didn't seem to matter as much when my mother was alive. She held us—and herself—together after my father died, after my sister Cathy's son Brian was born with severe handicaps, and Paul's son Tyler developed autism, and Johnny's son Andy committed suicide. I asked her once how she survived all the loss and the sadness. "You just do," she said. "You just keep going."

The going is harder now without her. "I know she's dead," said my brother Paul, a few weeks after the funeral. "I just can't believe she's gone."

My siblings and I are good about staying in touch. We still gather as a family to re-create those happy times in my grandparents' backyard. And we are successful to a degree. But we come to the festivities with heavier hearts now. I suppose the same was true for my grandparents, and for my Aunt Teresa and for all the members of that older generation who had lost many of their own loved ones by the time they came to preside over my childhood. But they disguised it well. Maybe it was their matter-of-fact natures that kept them going, or all the booze and bonhomie. I wish I had asked for their secret.

There have always been lots of children in our family. On my mother's side alone there are 22 grandchildren and, as of this writing, 62 great-grandchildren. Healthy babies were a birthright in our fertile Catholic clan. So we were completely unprepared for Brian.

My sister's middle son was born on February 18, 1987. He weighed six pounds, 12 ounces. By all accounts, Brian's was a normal delivery. But Brian wasn't a normal child. "Neurologically impaired" is how the doctors characterized his condition, which presented, in Brian's case, as a crippling trifecta of cerebral palsy, mental retardation and a seizure disorder. Experts ruled out physiological, metabolic and genetic explanations, venturing only that whatever had gone wrong had likely occurred at conception or early in the first trimester when the nervous system was still forming.

Was it really just reproductive bum luck? Was the river, or New Jersey somehow to blame? We'll probably never know. What my family did come to know is that Brian would never walk, sit up, speak or feed himself. And we were left to mourn, not death, but life, the life of the child that might have been.

My sister's life changed dramatically after Brian's diagnosis. Locating schools, therapists, doctors, shuttling Brian back and forth and caring for him became her fulltime job. Some days were devastating. Like the day she got the results of his first developmental evaluation, which showed that her 10-month-old son was functioning at a zero-to-three-month level. When she got home that afternoon, Cathy called our mother, in tears, to say, "I will never accept that this has happened."

But Cathy did come to accept what happened to Brian, better than I, better than any of us ever have. And somewhere in the painstaking, time-consuming, selfless business of caring for their severely handicapped child, my sister and brother-in-law entered what the nuns would have called a "state of grace." They stopped being angry. They stopped being sad. They stopped asking "Why?" They arrived at a place of peace.

The old Homelite plant is a strip mall these days. Still, I observe my own private moment of silence as we paddle past. The Rutgers Street

Bridge rises before us. The steel bascule version is gone now, replaced several years back by a new wider span with a different lift design. The new twin towers support a four-lane roadway that can rise between them like a dumbwaiter to let boats pass underneath. The decision to replace the old bascule bridge with another drawbridge implies that state officials expect big boat traffic to be back on this part of the Passaic some day. The PVSC skimmer crews surely appreciate the new working span, although their drawbridge problems were largely solved when the skimmer fleet relocated to its new headquarters at 403 River Road in North Arlington in 2009.

I miss the old Rutgers Street Bridge. Its black steel girders and giant concrete counterweight had a hulking grandeur. The new bridge just looks tacky, its twin towers like doublewide trailers on stilts. Perched four stories above the water, the trailer tops are painted a pale grey and white with brick-red trim. They dwarf the Arlington Diner, a fixture on the river's eastern shore. Even the tall steeple of Belleville's historic Dutch Reform Church on the opposite bank looks puny by comparison. There are 62 Revolutionary War veterans buried in its churchyard, along with Josiah Hornblower,[2] the famed Belleville engineer, entrepreneur and statesman whose mansion sat directly across Rutgers Street from the church. The graveyard is also the final resting place for many of James Hervey's Chinese laborers, who made their homes in Belleville and worshipped at the Dutch Reform Church.

The Passaic is not the rolling boil I remember from that day many years ago when I peered timidly through the metal grate of the old Rutgers Street Bridge. From here below, the water appears smooth and untroubled.

---

2 Josiah Hornblower was the British-born engineer who brought the first steam engine to America in 1753. The imported engine was used to pump water from the copper mines in North Arlington that were owned by Arendt Schuyler. Hornblower settled in Belleville, and became a successful merchant and respected civic leader. In 1785, he served as a New Jersey delegate to the Continental Congress.

Maybe the current and tide were making mischief that day. Or maybe I was projecting my own callow fears onto the river.

It takes six paddle strokes to clear the Rutgers Street Bridge. Six paddle strokes to leave North Arlington behind.

We are in Kearny now, a town of mostly Irish families known for its fish 'n chips restaurants. In 1979, the Argyle Fish & Chips Restaurant on Kearny Avenue (the Ridge Road of Kearny) got five stars from *The New York Times*. Waterfront trees, ablaze in the late afternoon sun, festoon the Kearny shoreline in red and gold. I have driven along the Kearny riverside many times, and viewed the river against the steel and concrete backdrop of McCarter Highway on the opposite bank. I've never seen the Kearny shoreline like this before, close up and from the river. I realize how much we east shore natives lost when Route 21 replaced the Passaic's western bank. I joke to Carl and Cathy that the state should bolt a giant mirror to the bulkhead in Belleville so that Kearny residents can enjoy the comparative lushness of their own riverside.

The tide is rushing out now. I can feel it pulling us downriver. Paddling is much easier, which is good because my arms are reaching their end-of-the-day heaviness. Carl was right. We will make it to Newark before dark, but just barely.

At the southernmost edge of Kearny, Carl spies a football bobbing close to shore. But not just any football. This is an official, leather NFL football and it can't have been in the river for long because it's in near-perfect condition. Some errant kick or toss probably put the pigskin in the drink. Carl rescues it with his usual relish, thrusting the ball above his head and grinning, as if he'd just scored the winning touchdown. He paddles over and awards me his damp prize, insisting that I give it to my brother Paul as a thank you for letting me stay at his house during all my many Passaic River research trips. (Which I do and Paul loves it. Carl is right again.)

We snap a few photos at Rapp's Boat Yard in Kearny, the last working boatyard on the Passaic and one more landmark whose existence I was totally unaware of before I ventured out onto the river. Rapp's looks a little run down. The white paint on its dockside building is peeling. There is a small fleet of decrepit watercraft beached along the shore. But six or so nice-looking powerboats and cabin cruisers are tied up at the long floating dock, so there must still be boaters looking for moorage on the Passaic.

We make one final stop before the take-out—in Harrison— to ogle a handsome red-tailed hawk perched above us on a metal fence post. The fence surrounds the parking lot of Marshall's department store on River Road. I paddle up slow and quiet until the bow of my Pungo meets the riprap. The hawk peers down at me but it doesn't budge. The bird looks as comfortable and as regal on that fence post as if it were surveying a windswept prairie from some dizzying promontory.

As wild things go, red-tails are pretty adaptable. They're omnivores for one, preying on small mammals, birds, reptiles and amphibians, and they're not above a little carrion now and then. Red-tailed hawks are flexible nesters too. They prefer to build in stands of large trees near the water, but a cliff or a ledge or even some kind of manmade structure will do. Red-tails have managed to live side-by-side with people, lots of people. I admire their ability to adapt and even thrive in the face of a habitat loss and destruction that by any measure could be considered catastrophic. Rank water, paved fields, more telephone poles than trees. The sturdy raptors with their piercing yellow eyes and tawny tails keep hunting and mating and nesting and raising their young. They keep going. Like my mother. Like the Passaic.

The end of our river trip is in sight now. We have traveled the Passaic for almost 75 miles, watching the degradation crescendo as we moved downstream. We are near our final take-out on the most toxic stretch of river, the dirtiest water. There are no more dams or waterfalls to negotiate, no boat races or fairs or wildlife sightings, and very little daylight left. There is just our little two-boat flotilla moving slowly down the smooth, dark, dirty Passaic, following the tide into Newark.

12 | THE SOURCE

THE FIRST TIME THE NAME "PASSAIC RIVER" APPEARS in a Hagstrom Atlas of New Jersey is on page 29 of the Morris/Sussex/Warren County edition. In the lower left-hand corner of this page two unnamed brooks approach each other—one from the south, the other from the west—and meet at a 90-degree angle. The stream that emerges from their confluence and flows briefly east before hooking south is clearly labeled *Passaic River.* Having just paddled most of the Passaic's navigable length, it was time I saw the river's source. With Hagstrom in hand, Phil Jaeger and I set out on a warm October afternoon in 2006 to visit the headwaters.

Phil is a retired high school math teacher from Cedar Grove, New Jersey, who has become an authority on Passaic River history. He is trim and compact with pale blue eyes, a prominent nose and close-cropped silver hair. He has a deep, resonant voice that commands attention, and he is a natural raconteur. The slideshow lectures on Passaic River history and other New Jerseyana that he delivers at libraries and community centers around northern New Jersey are entertaining affairs. Phil spices his capsule histories of the Passaic with photos of Colonial gristmills, native fish weirs and petroglyphs and the gravestone of Great Falls' leaper Sam Patch.[1]

According to Phil, who has been there once before, the birthplace of the Passaic is just a few miles from the historic Hilltop Church in Mendham. As its name implies, the church commands a rise near the center of town. Its tall white steeple is visible for miles. Phil makes a left out of the church's nearly empty parking lot. The bright mid-day sun makes it warm enough to roll down the car windows. The air smells of dried leaves and wood smoke and football and autumn.

Mendham, New Jersey is one of the most historic and picturesque communities in Morris County. The countryside is a semi-rural tapestry of rolling hills, winding roads, open fields, steeply wooded slopes and a preponderance of pristine streams and wetlands. Some 5,000 people reside in Mendham's inner Borough and outer Township. Most are aging, white-collar professionals, who are well educated and comfortably upper middle class. They live in large, single family homes on generous parcels of land. The six-square-mile Borough of Mendham is the hub for the 18-square-mile Township of Mendham, which surrounds it on three sides like a wrench. The Borough and Township are governed

---

1 Sam Patch, "The Yankee Leaper," was a young mill worker from Rhode Island. In September 1827, he stunned onlookers with a daring jump off the rock ledge at Great Falls in Paterson. The stunt launched Patch's brief career as a daredevil. With his signature, feet-first form, Patch knifed into the water from high atop bridges and factory walls and ship's masts around the northeast. In October 1829, he became the first person to successfully jump off Niagara Falls. By then, Sam Patch was full-on famous, the Evel Knievel of his day. Historian Paul E. Johnston, whose 2003 book, *Sam Patch: The Famous Jumper*, chronicles Patch's exploits, calls his subject "the first modern celebrity." Two weeks after his Niagara triumph—on Friday, November 13th—Patch attempted to jump into the Genesee River from Upper Falls in Rochester, New York. He is said to have visited a few Rochester bars before the stunt. Some of the 8,000 spectators who turned out to watch believe that Patch fell, rather than leapt, into the Genesee River that day. His body was discovered the following spring, miles downstream. Both shoulders had been dislocated, presumably from the impact of the fall. Cause of death: drowning. Patch was 22. He is buried in Charlotte Cemetery near Rochester.

independently, but they share a geologic, hydrologic and cultural heritage that is rich and unique.

The pride of Mendham is the six Historic Districts spread across the Township and Borough. Five are listed on the National Register of Historic Places; the sixth has been nominated. Many of Mendham's 140 historic buildings are clustered within the Borough's village center, site of the Hilltop Church and the Black Horse Inn & Tavern[2] where Phil and I had lunch before setting out on our field trip to the source. Mendham's other historic highlights include the 1914 Revival-style mansion and estate of Franklin Murphy, a former New Jersey Governor (1902-1905) and founder of Newark's Murphy Varnish Company; the Thomas J. Laughlin distillery (pre-1899), producer of a potent local applejack known as "Jersey Lightning;" and the old Rock-a-Bye-Baby Railroad[3] depot (1888).

In geologic terms, Mendham hails from the Highlands physiographic province. Both the Township and the Borough are poised at the southeastern edge of the Highlands, just south of the Wisconsin glacier's terminal moraine in nearby Morristown, and just west of the Piedmont. The bedrock here is all granite and gneisses, the signature hard rock types of the Highlands. This erosion-resistant foundation makes the terrain in

---

2 The Black Horse Inn at 1 West Main Street in Mendham has been in operation since 1740, which is the year Ebenezer Byram decided to go into the hospitality business. Byram, whose family settled Mendham, transformed his aging farmhouse and barn into an inn and a pub and began hosting the weary passengers traveling by stagecoach between Newark and Scranton, Pennsylvania.

3 The Rock-a-Bye Railroad was one of the many small lines that once crisscrossed New Jersey. Its official name was the Rockaway Valley Railroad. The nickname came from the rhythmic swaying of the railroad cars as Rock-a-Bye trains chugged along their 25-mile route, hauling peaches and passengers between the Jersey towns of Whitehouse Station and Watnong. Poor construction, a turn-of-the-century peach blight and a fire that destroyed the Watnong station doomed the Rock-a-Bye. The line went belly up in 1913. Some of its old rights-of-way survive today as park land or hiking trails around Morris County. There's one especially scenic stretch along the Black River, which is a tributary of the Passaic.

Mendham hilly. Some slopes in the area rise 800 feet above sea level. The highest point, near Horizon Drive, soars to 1,027 feet.

The rugged topography and the waterproof bedrock also create poor drainage conditions in the area. During rainstorms or snowmelt, the pitch of the landscape speeds water downhill along the surface. (Picture a paved parking lot on a hill.) Geologists call this "sheet flow." Water that pools long enough to soak into the soil can't penetrate too deeply into the solid bedrock below. The cracks and seams in the rocks closest to the surface provide some avenues for the movement of this groundwater, but at deeper levels these rock openings seal shut. Any precipitation that manages to percolate below ground doesn't get very far, and it doesn't stay buried for very long. Subterranean caches of water in Mendham tend to resurface as springs or seeps which coalesce to form wetlands or streams at lower elevations. Taken together, these physical dynamics, said geologist Otto Zapecza, favor "the accumulation of waters that feed secondary streams and tributaries of larger river systems." In other words, Mendham's Highlands-style geology makes it a spawning ground for rivers.

As we head east along tree-lined Talmadge Road, Phil recounts the serendipitous journey that led him to the Passaic. After his first career in corporate America—as a statistician with Shell Oil and then in market research for Lever Brothers—Phil spent 20 years teaching math and computer science at Millburn High School in Millburn, New Jersey. One Sunday afternoon in the early 1970s, while visiting his mother-in-law in western New Jersey, Phil came across an old Morris Canal bed about two blocks from his mother-in-law's home. The accidental find turned into a project, which blossomed over time into Phil's post-retirement career as a New Jersey historian.

Phil researched and photographed the Morris Canal system, walked most of its 102-mile length from Phillipsburg to Jersey City, and eventually served three terms as Director of the Canal Society of New Jersey. He began collecting vintage postcards and published postcard histories of Montclair and Cedar Grove. He juxtaposed the vintage postcards with his own current snapshots to create "photographic then-and-now" scenes, which he began incorporating into slideshows using a two-screen, two-projector setup. Before long he was tapping his old contacts to help him arrange dates and venues for the slideshow histories he was assembling.

Phil first crossed paths with the Passaic, quite literally, in 2001, right after his wife Jean retired from her teaching job. Phil was looking for something they could work on together, and one afternoon while he was leafing through a magazine feature on "fine country dining," it came to him: fine *county* dining. "Jean and I were going to eat our way through New Jersey's 21 counties, and try and uncover some history in the process," says Phil. On the day Phil and I visited the source, he and Jean were about two-thirds of the way through, having just completed Passaic County.

As they traveled around New Jersey, says Phil, "we kept coming across the Passaic River. It was here and there and everywhere, because it winds all over the damn place. So I said to Jean, why don't we go out to *where it starts?*"

Rivers start in all sorts of places on every continent on earth. A source might be a seeping spring, a weeping glacier, a mountain bog or lake or a cluster of ruts and gullies that gathers and funnels rainwater and snowmelt. The Hudson River rises at Lake Tear of the Clouds, a limpid glacial pool in New York's Adirondack Mountains where Vice President Teddy Roosevelt was camping in September 1901 when he got word that

President William McKinley's gunshot wounds were likely fatal.[4] The mighty Amazon, the world's largest river by volume,[5] first appears as a drizzle of ice melt on an 18,363-foot-high peak in the Peruvian Andes. The Seine starts with a cluster of springs on a high plateau northwest of Dijon in France's hallowed Cote d'Or wine region. Scientists recently discovered many large rivers that flow, during warmer periods, beneath the massive ice sheets in Antarctica.

The two conditions necessary for river formation are an excess of water and a slope to send that water flowing. Mendham, New Jersey has both, as well as a hard-rock geology that discourages pooling. Hydrologists consider Mendham—and most of the east coast for that matter—an "exorheic" region. The word (pronounced EX-o-ree-ik), comes from the Greek meaning to "flow from." In exorheic regions, precipitation exceeds evaporation. Exorheic regions like Mendham, like New Jersey, enjoy a net surplus of water.

This exorheic status is one of the reasons the area around Mendham—and the Garden State in general—once supported many farms. In addition to fertile soil, New Jersey gets plenty of rain. Most Jersey towns record an average of 43 to 47 inches of rain each year. The north-central part of the state is even wetter, measuring as much as 51 inches annually. Mendham, which lies in the damper north-central zone, averages 53.67 inches of rainfall a year. July and August are the soggiest months,

---

4 William McKinley, the 25th President of the United States, was shot twice on September 6, 1901 while he was greeting the public at the Pan-American Exposition in Buffalo, New York. Doctors initially believed that McKinley would recover from his wounds, which was why Vice President Theodore Roosevelt went ahead with plans to join his family for a camping trip at Lake Tear of the Clouds in New York State's Adirondack Mountains.

5 At 4,160 miles, Egypt's Nile River has long been considered the world's longest. In 2007, a team of Brazilian researchers challenged that claim to supremacy. Results from their 14-day mapping expedition adjusted the Amazon's length upwards, by 176 miles, to 4,225 miles. If confirmed, this finding will give the Amazon longest river bragging rights.

but rain falls fairly evenly from January through December.[6] The excess H_20 either soaks into the soil where it is absorbed by local plant life, or it flows, draining into the rills and brooks, which combine to form larger waterways that empty into oceans and seas. Exorheic regions produce rivers and send those rivers forth.

The opposite is true for "endorheic" regions. These are typically arid lands, such as the American Southwest, where evaporation exceeds precipitation. Networks of rivers and lakes in endorheic regions act like closed systems. Utah's Great Salt Lake is one example. Three rivers feed the Great Salt Lake: the Jordan River, the Weber River and the Bear River, which is the longest of the three, traveling some 350 miles from its source in Wyoming. River water that flows into an endorheic basin like the Great Salt Lake doesn't flow out. Neither does the rainwater that falls on the basin. The only way for water to leave an endorheic basin is by evaporating. The Jordan, Weber and Bear rivers are not destined for the sea. The Great Salt Lake is their terminus. Endorheic regions keep their rivers to themselves.

After we drive for a mile or so east along Talmadge Road, Phil turns left onto Colville Drive, a short connector street that links Talmadge Road and Lowery Lane like the rung of a ladder. We go right on Lowery Lane. At the end of this sleepy cul-de-sac, the pavement widens into a tear-drop shape and suburbia gives way to a grassy wetland with patches of low shrubs and stands of young alder, maple and oak trees.

Between the end of the Lowery Lane and the wetland there is a small manmade pond and the clear stream that was dammed to create it. Red and gold maple leaves float on the pond's dark surface. Right in front of us, at the end of the pond opposite the dam, sits a low concrete

---

6 Source: Office of the New Jersey climatologist.

overflow with a two-foot-wide weir at its center where the lake water spills through into a natural tailrace. I step carefully along the wall's slippery, moss-covered top and cross the weir to the other side, into the marsh.

Now it's Phil's turn. He shuffles across the slick wall to the edge of the weir and pauses there for a moment, steadying himself for the hop across the gap. As he prepares to push off, his back foot slips. He loses his balance and plunges his right leg into the tailrace. Phil stands motionless for a second, one leg on the dry bank, the other knee-deep in New Jersey. He pats his breast pocket to make sure his glasses are still there. And then he starts to laugh.

He is destined to get wet on Passaic River fieldtrips, Phil tells me while he sloshes out of the stream. When he visited the confluence of the Dead and Passaic Rivers, he slipped and fell while trying to take a picture. His camera survived the mishap, but he lost his reading glasses and his pants got so muddy that he banished them to the trunk of the car and drove home in his underwear.

Once in the marsh, Phil and I angle left, tromping through knee-high grass that is dry and brown now with autumn. In the soggier spots, I use the grassy hummocks like stepping stones to avoid the small, muddy pockets in between. Much of this lowland must be underwater in the spring.

In no time we come upon a clear, narrow stream that cuts straight across our path. The stream's channel is deep and distinct, but so obscured by the tall grass that we don't see it until we're right on top of it. We follow the stream flow to the left, north. Suddenly, Phil stops.

"This is it," he says, pointing to the map.

We have come about a quarter mile from the end of Lowery Lane. We can't see the car or the road or any houses.

I come up beside Phil.

We are standing in a grove of young trees. The canopy has shaded the grass here, and even though it is mid October the grass is still a bright green. Our presence disturbs the songbirds. They dart and chirp and rustle the brush at the edges of the clearing. It feels hidden here, like a secret garden where it is always spring.

To our right, from the south, comes the small rill we have been following through the grass. On our left, and almost perpendicular, is a twin trickle that arrives from the west in a wider, shallower and more open streambed. We are standing precisely at their juncture. The newborn Passaic is right here, right in front of us, bubbling off towards the east. It is larger than its two parent flows, but it isn't a river yet, not by any stretch. There are no bulkheads or factories or garbage. Just a stream, about three feet across with a shallow bed, cobbled in spots, and grassy banks littered with twigs and small branches.

Other sources[7] claim the western branch as the actual source of the Passaic. This rill forms near the junction of East Main and Orchard streets in Mendham Borough, and flows southeast for about a mile before it reaches the spot where Phil and I are now standing. I do not dispute their finding. It's just that I can't see a singular difference that would confer source status on the western stem over its ever so slightly shorter partner from the south. So for Phil and me, the Passaic, so pure and innocent, starts here in this sun-dappled glade at the end of Lowery Lane.

I wish I could freeze this moment, stop the baby river right here and now while it's still all fresh and clean and new, save it from its superfund fate. But rivers are destined to flow. They keep going. And the Passaic

---

7 In his 1974 book *The Passaic River: Past, Present, Future*, author Norman Brydon describes the headwaters of the Passaic as a "network of small streams rising in a marshy area on the eastern side of a steep, wooded slope just behind the center of Mendham. These unite to form a brook which runs along the edge of Mendham High School's athletic field, crosses a large area of open farm land, and proceeds towards Newark Bay..." Authors of the 2006 Master Plan for Mendham Borough concur.

is already on its way. It will course around the Great Swamp, through Millington Gorge and Great Piece Meadows, over Little Falls and Great Falls and past the cities and towns that will tarnish it with toxins and trash until this clear, bubbling beginning seems like a dream.

I can't help wondering what would have happened to the Passaic if the Wisconsin glacier hadn't blocked its original path through the Watchungs, if the river still flowed directly east to Newark Bay, bypassing Paterson and Newark and all their industries and chemicals and mistakes. It would be a shorter river, to be sure. Without that long detour north to Little Falls, the Passaic would measure 30 miles, not 86, and it would be much straighter too. Of course, without the Passaic running nearby, the Great Swamp and Great Piece Meadows and the other damp remains of Glacial Lake Passaic would likely have dried up and given way to forest and development by now. There would still be a Little Falls, and a Great Falls too, but it would be the Pompton River, not the Passaic, rushing over them. Towns in the Lower Valley would still be river towns. But Otto Zapecza would have been hooking Pompton River carp up in Garfield. And I would have grown up a stranger to the Passaic, with a whole new river—the Pompton—to fear.

I don't know that a shorter, straighter Passaic River would have turned out any cleaner. The river would still cut through industrial Elizabeth on its way east to Newark Bay. Besides, where rivers go, civilization tends to follow. So it is with today's Passaic. So it likely would have been with the what-if, Mendham-to-Newark express river that I imagine.

Mendham gives rise to three of New Jersey's major river systems: the Passaic, the Whippany and the North Branch of the Raritan, which is the state's second largest river. All three trend downhill from their starting point. The Passaic drops 600 feet between Mendham and sea level Newark Bay.

Building a river system is an orderly, incremental process. Small headwater brooks unite to form streams, which merge with other larger streams and so on down the line until a full-fledged river emerges. The resulting web of water, from the smallest starter rivulet to the final majestic river, creates a drainage network, or watershed. The tight-knit family of waterways fits together like the branches on a tree.

Branching patterns are common in nature. Trees have them. Veins and arteries have them too. The joining process is not random. There are rules and a nomenclature. The "successive merging is highly organized," writes hydrologist Luna Leopold in *Water: A Primer*. "It is one of the many aspects of dynamic equilibrium that is maintained within a river system." For example, every stream within a watershed is classified. Headwater streams are first-order streams; that is, streams without tributaries. When two first-order streams merge, they form a second-order stream. The confluence of second-order streams produces third-order streams and so on. The end point is the river that gives the watershed its name.

A watershed's design is not accidental. Certain constants define its architecture. Mathematical relationships exist between the number of streams in a given system, their length and their order. For example, the number of streams is inversely proportional to their order; so a watershed will always have more lower than higher order streams— approximately four times more. Stream length, on the other hand, is directly proportional to stream order, which means that third order streams are always longer than second order streams, which are longer than first order streams.

These relationships hold true for other networked systems as well. In fact, the relationships among number of branches, length and order are similar for trees and watersheds. The form of these systems serves their function. A tree must keep its leaves exposed to the sun. A river system must collect water from the land and transport it to the sea. "Branching patterns in trees, rivers, blood vessels in animal tissue, and other natural networks...are designed for efficiency and stability," writes Luna Leopold. A branching design makes their tasks easier.

Mendham's web of clean, swift waterways made it a logical site for industry, and Mendham did dance, albeit briefly, with the beast. A water-powered boomlet in the late 18th century brought mills, stills and iron forges to the area. By the early 1800s, those pioneering industries were joined by carpet, glass and rug-weaving factories, lime kilns, cement quarries, tanneries, nail-making shops and mica, lead and copper mines. Ice became a hot business for a time, with mill owners carving blocks from their frozen millponds for sale. Fortunately, for the area's headwater streams, this industrial period was short-lived. The bubble burst when local mills and factories failed to keep pace with the larger, steam-powered operations that flourished after the Civil War. By the late 19th Century, industry in Mendham had given way to farming. Apples were the main crop; that is, until 1920 when Prohibition put the kibosh on production of the local hard cider known as Jersey Lightning. There is no industry to speak of in modern-day Mendham. But there is growing pressure to develop previously open space.

Headwater streams in Mendham provide potable water for many Morris County residents. In recognition of their public health and ecological importance, they are classified "Category One" by the New Jersey Department of Environmental Protection.[8] This designation grants

8 According to the New Jersey Department of Environmental Protection's Water Monitoring and Standards, Category One waterways are designated "for protection from measurable changes in water quality characteristics because of their clarity, color, scenic setting, other characteristics of aesthetic value, exceptional ecological significance, exceptional recreational significance, exceptional water supply significance, or exceptional fisheries resource(s)." At present, the state DEP has granted Category One status to nearly 4,000 miles of streams and more than 10,000 acres of lakes and reservoirs in New Jersey.

Mendham's waterways special protections under state law. Buffer zones of 300 feet must be maintained between headwater streams and major development projects. Any new subdivision has to have conservation easements as a way to safeguard water quality. These easements, a kind of environmental restraining order, take the form of corridors or other open spaces that keep human development at arm's length from sensitive natural habitats.

Because residents of Mendham live in the midst of so many protected waterways, they enjoy a luxury that is rare in the state of New Jersey: unpolluted, unobstructed brooks and streams. Mendham water is so clean that most area households draw their water directly from local wells. Many Mendham brooks actually support trout populations.

The proximity to unspoiled water also makes Mendhamites environmental stewards. It appears that they take their stewardship role seriously. In 1993, the community overwhelmingly approved an Open Space Trust Referendum, which devotes a portion of local property taxes to keep open space open. When a company that trains seeing-eye dogs recently moved to nearby Chester, New Jersey, Mendham's Open Space Trust fund helped the community buy the multi-acre, seeing-eye property, which will now remain undeveloped. "[Mendhamites] are big on protecting open space," said Russ Heiney, the Construction and Zoning Official for Mendham Township.

The desire on the part of local government and the local citizenry to preserve open space is good news for the health of Mendham's headwaters. But authors of the most recent Borough (2006) and Township (2002) master plans warned that the surge in building over the last two decades has nevertheless threatened protected waterways. They reminded Mendham's elected officials of their "responsibility" to adopt strict land use regulations, and they urged citizens to do their part by limiting the use of lawn fertilizers, by fixing failing septic systems and by considering the impact of everyday household products such as non-phosphate-free detergents on the fragile ecology of the Passaic, Whippany and Raritan river headwaters.[9]

At roughly 200 acres, the quiet wooded marsh that nurses the Passaic is one of Mendham's larger undeveloped areas. It sits within a triangle of green that is bounded by the Lowery Lane community to the southwest, quiet Corey Lane to the southeast, and to the north by Tempe Wick Road[10], a pretty east-west thruway that begins as Watchung Avenue in Chatham to the east and ends 12 miles later at Main Street in Mendham, just a half mile northeast of the Black Horse Inn. Two large parks lie just north and east of these borders. The smaller of the two, Lewis Morris County Park, is named for New Jersey's first Governor. The larger Morristown National Historical Park, which bestrides Tempe Wick Road, commemorates the place where George Washington and his troops survived two frigid winters.[11]

---

9 Mendham Township's 2002 Master Plan states that: "The protection of public health and environmental quality are the fundamental responsibilities of Mendham Township, its boards and public officials. To this end, the protection of groundwater and surface water quality and quantity should be the sine qua non of Township future land use planning policies."

10 The prosperous Wick family owned a farm in what is now Morristown, New Jersey. More than 13,000 Continental soldiers camped on their property during the winter of 1779-80. Temperance "Tempe" Wick was the family's spirited teenage daughter. Legend has it that she was out riding one day when she encountered a band of deserters who tried to steal Colonel, her fine horse. Tempe escaped when one of the soldiers momentarily let go of Colonel's bridle. She galloped home and, realizing that the soldiers would come looking for the horse, led Colonel into a guest room on the second floor of the farmhouse and hid him there for three weeks. Tempe Wick is buried in Morristown's Evergreen Cemetery. The Wick House, where she hid her beloved horse, is preserved in the Jockey Hollow section of the Morristown National Historic Park.

11 The first Morristown encampment, during the winter of 1777, followed George Washington's historic Christmas crossing of the Delaware River and subsequent victories in Trenton and Princeton. The Continental Army spent the winter of 1778 in Valley Forge, Pennsylvania. But Washington's men returned to Morristown the following winter (1779-80). Washington chose the Morristown site for its clean streams and for the natural protection provided by the nearby Watchung Mountains and Great Swamp. But the price for water and security was bitter cold. The winter of 1779-80 was the worst of the 18th Century. Seven blizzards battered the camp in the month of December alone.

The two parks inoculate the tract's northern and eastern flanks against development. According to Russ Heiney, the green zone itself is owned by the Girls Scouts of America. "They use it during the summer for camping," he said. "They've probably had it for the last 50 years." The property is zoned residential. As far as Heiney knows, the Girl Scouts have no plans to sell or develop the land. But even if they did, says Heiney, it's unlikely that houses would threaten the Passaic. "We're all built out," said Mendham Township's zoning officer. "We have one large tract that has approval for about 10-12 homes. And that's it." No more houses in Mendham.

The Passaic's first brush with human endeavor occurs soon after the infant river leaves its 200-acre incubator. A mile downstream from the source, at the intersection of Tempe Wick Road and Leddell Road, the Passaic is impounded behind an old stone dam. The dam and the millpond it sustains were built by William Leddell to power the gristmill and sawmill that he erected on the site in the 1770s. (The Leddell property is on both state and national historic registers.) On our way out of Mendham that afternoon, Phil takes me by the old Leddell place. It's a short and pretty jaunt through the woods.

We park Phil's sedan on the shoulder, just east of where Leddell Road joins Tempe Wick. We walk the short distance back to the spot where Tempe Wick crosses the Passaic, just below the Leddell dam. Though it is unnamed, the short, low archway we are standing on, with its chest-high walls of blonde stone, is the first Passaic River bridge. From here, the bridge at Tempe Wick Road, we have a clear view upstream, some 200 feet, to the dam and the pond and what remains of William Leddell's mill.[12]

---

12 According to John W. Rae, author of *The Mendhams*, a photographic history of the area, there were two mills on the Leddell site. The first, more primitive mill was replaced at some point, and a second, stronger dam was built to serve the new operation. The Leddell mill ground grain and flour for the 13,000-plus Continental troops who camped at Jockey

The mill is a ruin now. The few remnants of its stone foundation are being overrun by a march of vines and maple saplings. By contrast, the dark stone dam, about six feet high and some 40 feet across, is entirely intact. Autumn-tinted trees frame the long narrow millpond behind the dam. The pond is low this time of year. No water is spilling over the flat rock slab that caps the dam's top. But a small outlet feeds the Passaic below, which burbles along through the woods, winding south past Jockey Hollow[13] and under Route 287 towards the Great Swamp, some five miles downstream.

Before Phil and I return to the car, I snatch the prettiest leaf I can find, a blazing red and gold maple leaf, from the bridge at Tempe Wick Road and drop it into the river below. The leaf flutters down slowly, like a moth drawn to the flames of sunlight on the water. A sudden breeze threatens to blow the leaf off course and strand it on the riverbank, but gravity prevails and my red and gold offering finally settles on the Passaic and begins its pirouette downstream.

As I follow the leaf's progress from the Tempe Wick bridge, I'm reminded of something Carol Johnston told me as we toured the Ironbound. She was talking about the tendency of western civilization to discount animist beliefs. As a Catholic nun who has studied cosmology, the idea that a spirit, a soul, animates all things— from plants and animals to rivers and seas— resonates with Johnston. She regrets the extent to which the idea

---

Hollow during the bitter winter of 1779-80. The stone ruins visible today are the remains of the second mill, which may have been in operation as late as 1905. The Leddell home, also on Tempe Wick Road, was constructed of stones that were salvaged from the campfire rings kept by troops at Jockey Hollow.

13 At Jockey Hollow Encampment, the southern arm of Morristown National Historical Park, Continental Army soldiers cut down acres of trees to build more than 1,000 shelters where they hunkered down for winter. The "log huts" were arranged in orderly rows and built to Washington's own precise specifications. Each hut measured 14 by 15 feet and was shared by 12 enlisted men. Officers' quarters were slightly larger and housed between one and four. Washington's original "log-house city" is gone, but log huts have been reconstructed at the original Jockey Hollow site.

and its practitioners have been marginalized. She is sorry that humankind has grown so estranged from the natural world. "We have no sense of deep connection to a place," she said, on the day we drove along the Ironbound's narrow, cramped streets. "We sort of live on top of a place." When we are forced, as we are now, to face a place-related problem as complex as healing a ruined river, we find ourselves at a bit of a loss. "We have a lot of information," said Johnston, "but no insight."

13 THE END

On a cold, blustery day in November 2007, a full year after my Passaic River paddle and hike to the headwaters, I pulled out my list of rivers. I sat on my couch, where I could watch the whitecaps scudding across Vashon Island's Fern Cove, and reviewed the entries. They unleashed the familiar torrent of sensations: the sting of rain against my cheeks as I stared out at the congested, hurricane-besieged Hudson, the relentless surge of the Nisqually as Kate and I grabbed for riverside branches to try to pull our kayaks back upstream, the ink black stillness of the Rio Negro at dawn. The rivers on my list are old friends, each one stirring up vivid memories. But they didn't feel like family. Not like the Passaic. I feared my hometown river as a child, loathed it through my adolescence, and abandoned it at the first blush of adulthood, but the Passaic stayed with me. It was always there at the foot of our hill, in the back of my mind. It was part of me. Memories of the other rivers on my list were often fonder. But only the Passaic River felt like mine. I took out my silver Cross pen, the one Kate gave me for my 48th birthday, and added the Passaic River to my list.

I spent Christmas in New Jersey that year. One Sunday afternoon during my visit, I took a group of my nieces and nephews to see Great

Falls. Most of them live in New Jersey, or nearby in Pennsylvania. We were an even dozen: three nieces, six nephews, my brothers Joe and Paul, and me. I was the only one who had ever seen the falls.

The outing was a post script to our longstanding post-holiday tradition, a day-long family extravaganza that begins at 10am with ice skating at South Mountain Arena in South Orange, moves to my brother Paul's house for pizza, proceeds to a local theater for some holiday flick, and ends about 12 hours later with a chili dinner at cousin Nancy's house in Cedar Grove. "The tradition," as we call it, started in 1994 when my brother Joe and I went in together on a Christmas present for Cathy's sons Johnny and Paul. They were nine and four, respectively, at the time. We bought them each a pair of hockey skates and took them skating at South Mountain on the day after Christmas. A group of cousins came along that first year. Most of them have been coming ever since. I missed one year with a sinus infection. But Johnny, Paul and Joe have never skipped a holiday skate. Johnny is 27 now; Paul is 22. These days, they're the ones helping the little kids around the rink.

We scheduled our Great Falls tour for the day after skating. We started in the Visitor's Center parking lot. The small red wood building is just across McBride Avenue from the statue of Alexander Hamilton, where I rendezvoused with Ella Filippone two years earlier. We paused near Hamilton so I could get a photo of the group with the falls and its sandstone vault and the red brick power generating station in the background. Then I led my charges along the same path that Ella had taken me: up McBride Avenue, past the low dam above the falls and across the green footbridge to the killer overlook at the very end of Mary Ellen Kramer Park.

The river was magnificent that day, higher than I had ever seen it. The spray from the falls was visible before we even got close to the dam. It hung above the river like a diaphanous white veil against the winter blue sky. The green footbridge was dripping with condensation, and so were we by the time we crossed over it into Mary Ellen Kramer Park. The kids ran ahead, delighting in the river's smoke and thunder, and in each other. My brother Paul and I hung back, debating the health risks of inhaling aerosolized Passaic. By the time we all met up at the final overlook everyone was grinning and yelling to be heard above the water's roar. My brothers, nieces and nephews squeezed onto the small viewing ledge, just 20 short feet from the vertical torrent. I took a picture of them there, smiling in the sun and the cold, leaning against Ella's despised iron fence. The surging wall of river was at their backs.

Behold the mighty Passaic, I thought as I clicked the shutter. Remember it.

Much has changed for the river since that day, most of it for the better. The National Park Service is studying plans for the Great Falls Park upgrade and for a proposed "Passaic River Blueway," a water trail for paddlers that could install a staircase and adjacent boat slide into that steep muddy rut in the ravine below little Falls where Carl, Cathy and I almost lost the blue kayak. Newark has an energetic, Passaic-friendly administration with Mayor Corey Booker and popular city planner Damon Rich. The state has an agreement with Occidental Chemical, the corporation that owns the old Diamond Shamrock site, and an Occidental-sponsored project is underway to determine how best to dredge dioxin hotspots from the Passaic.

The project's Phase I cleanup targets the hottest spot of all, the 40,000 cubic yards of dioxin-rich river sediment at the Diamond Shamrock Superfund Site. Phase II will tackle the other 160,000 cubic yards of contaminated sediment that is concentrated along the shoreline areas just upstream and downstream from Lister Avenue.

Engineers responsible for the cleanup are still puzzling over the best ways to dredge and dispose of all that poison muck. Even the seemingly simple parts of the process present unanticipated challenges. They had hoped, for example, to use vacuum-like hydraulic dredges to quickly suck up sediment. But crews were finding so much junk in the river bottom—cars, old tires, lots of wood—that they will likely go with slower, clumsier clamshell dredges instead.

Once the contaminated sediments are topside, the water associated with them will be separated out, scrubbed and sent back into the river. The toxic mud that remains will be sealed up and transported by barge or boxcar to an off-site facility for incineration. But the dioxin won't die there. The toxic ash that survives the incineration process will have to be secured somewhere, probably in some landfill on the outskirts of some small middle-of-nowhere North American town. Someplace like Eunice, New Mexico, which has already agreed to babysit 1.3 million pounds of PCBs from the ongoing Hudson River cleanup.

The Phase I removal was supposed to begin in the spring of 2011 and last for two-and-a-half years. That timetable has been delayed. The Phase II schedule is yet to be determined. The U.S. Environmental Protection Agency considers the Passaic River dredging project a "Non-Time-Critical Removal Action." The upside of this NTCRA designation is that the delicate, dangerous dioxin removal won't be a rush job. The downside is that it could drag on forever, which would be a terrible failing, since Lister Avenue is already the oldest unresolved Superfund site in the country.

It's hard to say when the Lister Avenue remediation will be done. Every Passaic River advocate wants to believe that it could happen in his or her lifetime. "There *will* be a comprehensive cleanup of this river," vowed Lisa Jackson in 2008 at what would be her last appearance at the

Passaic River Symposium as New Jersey's DEP commissioner. In making such a bold pronouncement Jackson, now chief of the federal EPA, was walking her own desire line that day. In addition to removing and disposing of contaminated sediments, the EPA-mandated Passaic River cleanup calls for restoring habitat along nearly a mile of affected shoreline, which means that one day—at least a decade and probably billions of dollars from now—nature, with the help of Coastal Restoration specialists like Carl Alderson, could reclaim some of the most poisonous parts of the Passaic.

"The history of Newark is held in the river," Carol Johnston told the crowd at the 2008 Symposium. As is her habit, Johnston spoke a simple, yet profound truth: Everything we've done and haven't done, everything we are and are not is mirrored back to us from those bullied brackish waters.

Johnston's longstanding effort with the Ironbound Community Corporation to create a park along the Passaic waterfront has found a receptive audience in the Booker administration. She telegraphed genuine excitement as she shared the dream of one elderly Ironbound resident, who imagined "a green park along a blue river, instead of all that dirt." Johnston seemed hopeful that the dream of a river park was within reach, that the venerable old city of Newark might just reclaim the inventive, can-do mantle of its early days. "Our goal is to create a mile of green along the river," said Johnston. "Our strategy is to *begin*."

Andy Willner retired as executive director of the NY/NJ Baykeeper's Association in April 2008. He doesn't go to the Passaic River conferences anymore. He no longer takes newbies like me on Passaic River boat tours. But when we met for breakfast a few days after the '08 Symposium, at the Kenilworth Diner in Kenilworth, it was obvious that Andy was still following Passaic River goings-on.

He remains a hardcore skeptic about cleanups that involve public-private partnerships, especially when the private partner is the likes of Occidental Chemical, the same company, he reminds me, that was responsible for dioxin contamination at New York's infamous Love Canal. Andy's approach, for which he has lobbied publicly, passionately and so far unsuccessfully, is to refill the federal government's depleted Superfund coffers, spend that money on a comprehensive federally-supervised cleanup, then sue Occidental and other responsible parties for triple damages. "It's a minority position now," he allowed. "But there's no such thing as cooperative. In every cooperative [government-industry] effort around the country, the environment has been a victim."

The Passaic River has played the victim for a long, long time. "It's a poor people's river, at least in its most polluted run," said Andy. "A gritty, urban, blue-collar river that runs through minority communities. The attitude from government is that this is not a river worth saving." In Andy's view, the only way to turn that attitude around is for the people to take back what is rightfully theirs. "There's a right to clean water and a right to clean fish," he said. "There's a *law*. The more people assert those rights, the more impetus there is to do something."

Andy did find cause for optimism. "My last trip on the lower Passaic I saw people on jet skis, people in small boats, people in kayaks," he said. "It was just astounding." He believes that with an engaged community and a steadfast, activist government "we could probably have edible striped bass in the Passaic River and Newark Bay in 20 years. That's an instant in time for the river. It's a long time for people.

"At Baykeeper, people would say, 'we don't have the money to be engaged in this. We don't have the staff.' It turns out that we did have 20 years to throw at the issue, and that's what it's going to take. It's going to take long-term engagement by the public to make sure that the government does its job."

We have to think in geologic time.

Brian, my handicapped nephew, didn't go with us to Great Falls that day. A severe bout of pneumonia left him— and his parents— largely homebound. Brian died in January 2008, just a few months after my breakfast with Andy Willner, and one month shy of his 21st birthday.

In an effort to make some sense of Brian's life I reached out to Jim Albrecht, a friend and former colleague who is writing a book about science and religion. Jim is a philosopher, a romantic, a gifted intellectual and a true believer who applies a scientific rigor to the mysteries of faith. He is the only practicing Catholic I know who still goes to confession regularly. He once told me that if people took that sacrament seriously there would be no need for psychotherapy.

What, I asked Jim in an email, was the purpose of a life like Brian's? The point? Hadn't "God" wasted a perfectly good soul by assigning it such an imperfect body? A body that couldn't feed itself, let alone the hungry. That couldn't even think, or say a proper prayer.

Jim responded, by email, the very next day.

"I guess I would take two different routes to answer that one," he began. "First, this is a kind of variety of the 'problem of evil' question. Why does bad shit happen? And I guess a very old answer has been that, in order to make a real world, the stakes have to be real. The world has to be authentic. So you have this sense that cuts across many cultures and religions that the world is a thing of wonder, but a ruined wonder. In Christianity we call this 'the fall,' the notion that human sin has caused suffering and death to echo through history. The full meaning of any act or life is only revealed in the full history of the world. So the full meaning of your nephew's life can only be explained in terms of the whole history of an authentic universe and his role in it, most particularly in the lives of the people who were nearest to him and most affected by him, spreading out from there across all time.

"The other way I would try to answer that question is to say that we have a very special perspective in that we are living in time and in matter. We tend to attach all importance to this little glimpse we have of existence. We assume [our glimpse] is normative of the way things are and that it gives us a good indication of the scope and breadth and full nature of existence. But isn't it just as reasonable to assume that our perspective, while unique, is a rather small one? Your nephew may be involved in a great adventure now, something that makes what you think of as his life seem about as important to him as you would consider your own gestation, [when you were] trapped inside a space the size of your own body with no freedom and no thoughts for nine months. It would be an unthinkable hell if you were subjected to it today, but at the time, you didn't seem to mind.

"There is a principle revealed by quantum mechanics that an object cannot be properly understood in reference only to its past: you have to know its future as well. Similarly, in order to know the significance of your nephew's life, you have to know the full history, meaning the future. Until you do, you don't really have the basis to make a judgment."

As I suspected, there was no definitive answer to my cosmic thumbsucker. Still, I was comforted by Jim's response. I envy his faith in a life and a meaning beyond the here and now. I hope he's right. I hope Brian— and the Passaic too— is off on some "great adventure" whose import, whose *point*, will only become manifest in the fullness of time.

On my last trip back to New Jersey, I did something I almost never do: I flew into Newark in the daytime, instead of my typical after dark arrival.

The pilot approached the airport from the northwest, which took us over the Harrison and Newark reaches of the Passaic, over the big gooseneck of river that curls around the Ironbound.

Looking down, I saw the brown river in the brown land. The view reminded me of the elderly Newark resident whom Carol Johnston had quoted in her Passaic River Symposium comments. That old woman must have looked at all this "dirt" for a lot of years. But instead of despairing, she imagined "a green park along a blue river." She looked at the Passaic and imagined something beautiful. All the loss hadn't driven the hope from her heart.

After 30 years of Passaic-free living, I dove into my hometown river because I wanted a fuller picture of the Passaic. I wanted to confront my childhood fear of the river and replace it with something that I could take pride in and maybe even love. I learned that I was right to be afraid of the river, and smart to recoil from its carcinogenic filth. But blaming the Passaic for what it had become was an uninformed and misguided reaction. The dioxin, the PCBs, the heavy metals and sewage and garbage. The river didn't do any of that. We did.

I was afraid of the Passaic as a child. Now, as a grownup flying over the river I have traveled down, I am afraid *for* it.

I read an account once about Dan Simberloff[1], a reknowned evolutionary biologist who did his doctoral research on several small mangrove islands in the Florida Keys. Like most field researchers, Simberloff came to know and love his study area. He continued to visit the Keys even after

---

1 The results of Dan Simberloff's doctoral research, which appeared as three separate papers in the 1969 and 1970 editions of the scientific journal *Ecology*, represented a landmark in the study of island biogeography. Simberloff and his work figure prominently in author David Quammen's 1996 book titled *The Song of the Dodo: Island Biogeography In an Age of Extinctions*.

his doctoral fieldwork was done. Each time he went, he noticed a little more development, a little less nature. On one trip, in the mid-1970s, Simberloff was startled to see that Ohio Key, which was adjacent to two of his mangrove islands, had been clearcut to make way for the Sunshine Key RV Resort. Simberloff pulled off the road a few miles past the now lifeless Ohio Key and he sat in his car and he cried. After that, Dan Simberloff never went back to the Florida Keys.

I am haunted by this image of Simberloff, by all accounts a blunt, exacting man of science, sitting all alone in his car on a hot sunny Florida day, surrounded by sparkling emerald green waters, weeping for the bulldozed mangroves and palm trees and bugs and birds and frogs and shrimp and fish and life of Ohio Key. I grew up in New Jersey. I know the feeling.

It's dark by the time Carl, Cathy and I stow all the gear and strap the wheelies onto the two kayaks. It's a moonless autumn night—clear and crisp. The gentle lapping of the Passaic gives way to the urgent shriek of traffic on Raymond Boulevard in Newark. The continuous stream of cars and trucks on this three-lane highway makes our crossing harrowing, but we manage our last portage safely and head up Somme Street towards Market Street, rolling the boats along behind us. It's well past 7pm on a Sunday night, but the neighborhood is animated. As we near Market Street we pass the Riverbank Park soccer field, its lights ablaze for a high school girls' match that's in full swing. The parents and friends crowding the sidelines erupt with every play. Near the iron fence that wraps the field, two adolescent brothers idly dribble a basketball. The younger one shouts out to Carl as we pass. "Those canoes?" he yells. "Kayaks, stupid," says the older boy, shoving the ball into his kid brother's chest.

We had planned to toast the end of our river trip over dinner at one of the Ironbound's famous Portuguese restaurants. But it's late and we still have to drive back to the put-in to pick up my rental car. We decide, reluctantly, to postpone the celebration and call it a night. Before we drive off, we snap a few final pictures of each other, and of the boats, secured once again on the roof of Carl's Camry. Carl takes one last close-up of the three of us, squeezed together, with the blue Perception hovering in the background. With his long arm, he stretches his camera out and clicks. The picture is dark and grainy and slightly out of focus. We have big grins and bright eyes and we look happy and tired and proud.

I used to think it would take a great leader to rescue the Passaic, some kind of mad river czar with a fat budget and an iron will who could strong-arm polluters and preside over a genuine restoration. I realize now that there will never be a single savior, nor could there be. Rescuing the river requires more than one agency chief or elected official or nonprofit zealot. The kind of comprehensive change that the Passaic River needs can only happen when the Passaic River community demands it, loudly, consistently and for a very long time. It has to come from us.

People often ask me how they can help the river. I don't have a stock answer. There is no 12-step program for the Passaic. But there is this: all of us should find some way to get involved, any way. Take your kids for a walk along the river, write a check to one of the river advocacy groups, volunteer, complain to town hall about the garbage on the banks, complain to the neighbors, organize a cleanup party, a sit-in, write your Congressional representatives, your local newspaper, write a book. Be the squeaky wheel.

Do it for your children, for your health, for your property values, for the tax write-off, for show, for the beauty of nature, for the helluva it, for that feeling you get when you've done something good. Do it for the Passaic. Just do it. And do it soon.

Sixty years after Aldo Leopold's *Sand County Almanac*, human technology and gadgetry have grown even more sophisticated. We have artificial joints and predator drones and nanobots that seek out and destroy cancer cells. But in all that time, we have not mastered the art of living on the land without spoiling it. In truth, we haven't really tried.

And what will become of us if we really do manage to murder nature? The Swiss psychiatrist Carl Jung believed that "we all need nourishment for our psyche," and that it is "impossible" to find that nourishment "without a patch of green or a blossoming tree."

"We need a relationship with nature," Jung concluded. Without it, "I would not be a human being."

Sometimes I imagine a dark future for the planet, a future where all rivers are polluted like the lower Passaic, all seashores are dead zones, all mountaintops have been mined away. A world that resembles the landfill outside Eunice, New Mexico, or the grey mound at 80 Lister Avenue. A world without nature. A ruined wonder.

Rather than dwell on my bleak little vision I prefer to believe that we'll wake up in time and choose beauty. Watching residents of Brooklyn, New York begin to reclaim their long-abused Gowanus Canal[2]—another federal Superfund site—makes me hopeful that I'm right.

Ironbound activist Nancy Zak called the Passaic a "wild thing" that courses through a scarred and crowded land. She called the Passaic "powerful." She wasn't talking about the river's raging floods, mighty as they are. She was talking about the Passaic's power to stir souls, even in the darkest

---

2 The 1.8-mile-long Gowanus Canal, completed in the 1860s, passes through several Brooklyn, New York neighborhoods before emptying into Upper New York Bay. Like the Passaic River, this one-time tidal creek became a busy industrial thoroughfare and a sink for contaminants from waterfront coal yards, tanneries, ink and paint factories and more. The Gowanus was placed on the Environmental Protection Agency's Superfund National Priorities List in March 2010.

part of Newark, where the river's banks are rusting steel and its sediments ooze carcinogens. In the end, that power, more than anything, will be what saves the Passaic.

None of us can know the river's future. But we can help shape it. A clean blue river running through a cleaner, greener land doesn't seem like such a crazy dream, does it? The Passaic is still alive. People still care. The story isn't over. It continues on, like the river itself, in a long crooked arc through swamps and meadows, past cities and towns, over mountains of basalt and great caches of poison, in search of what we all long for: a happy ending.

Bibliography

Allen, Robert. *The Dioxin War: Truth and Lies About a Perfect Poison.* London, England: Pluto Press. 2004.

Baldwin, Douglas. *The Great Falls (Paterson, NJ): Representation in Art and Literature 1790-1845.* Montclair State University Master's Thesis. 2005.

Brydon, Norman F. *The Passaic River: Past, Present, Future.* New Brunswick, New Jersey: Rutgers University Press. 1974.

Cawley, James and Margaret. *Exploring the Little Rivers of New Jersey.* New Brunswick, New Jersey: Rutgers University Press. 1961.

Cleveland, Cutler. 2006. *History of the U.S. Fish and Wildlife Service National Wildlife Refuge System.* The Encyclopedia of Earth. http://www.eoearth.org/article/History_of_the_U.S._Fish_and_Wildlife_Service_National_Wildlife_Refuge_System#gen1. November 2011.

Cole, Gerald A. *Textbook of Limnology.* Saint Louis, Missouri: The C.V. Mosby Company. 1975.

Cunningham, John T. *Newark.* Newark, New Jersey: The New Jersey Historical Society. 1966.

Diamond Shamrock Chemicals Company vs. The Aetna Casualty & Surety Company, 609 A.2d 440 (Superior Court of New Jersey, Appellate Division 1992.)

Fariello, Leonardo A. *A Place Called Whippany: The History and Contemporary Times of Hanover Township, NJ*. Whippany, New Jersey: Len Sunchild Publishing Company. 2006.

Fietsam, Jr., Robert C., Judy Belleville and Jack Le Chien. *Belleville, 1814-1914*. Charleston, South Carolina: Arcadia Publishing. 2004.

Galishoff, Stuart. *Newark: The Nation's Unhealthiest City, 1832-1985*. New Brunswick, New Jersey: Rutgers University Press. 1975.

Gertler, Edward. *Garden State Canoeing: A Paddler's Guide to New Jersey*. Silver Spring, Maryland: The Seneca Press. 2002.

Gordon, Michael. *Protecting the Passaic: A Call to Citizen Action*. Seton Hall Law Review. Volume 29, No. 1. 1998.

Hamilton, Alexander. *Writings: Report on the Subject of Manufactures*. New York, New York: Columbia University Press. 1979.

Ianuzzi, Thomas, J., David F. Ludwig, Jason C. Kinnell, Jennifer M. Wallin, William H. Desvousges and Richard W. Dunford. *A Common Tragedy: History of an Urban River*. Amherst, Massachusetts: Amherst Scientific Publishers. 2002.

Jefferson, Thomas. *Notes on the State of Virginia*. Chapel Hill, North Carolina: University of North Carolina Press. 1954.

Johnston, James P. *New Jersey: History of Ingenuity and Industry*. United States of America: Windsor Publications, Inc. 1987.

Kraft, Herbert C. *The Lenape Indians of New Jersey*. South Orange, New Jersey: Seton Hall University Museum. 1987.

Kummel, Henry B. *The Problem of the Passaic Meadows*. Reports of the Department of Conservation and Development, State of New Jersey. Trenton, New Jersey: Beers Press. 1919.

Leopold, Aldo. *A Sand County Almanac.* Oxford, England: Oxford University Press. 1949.

Leopold, Luna B. *A View of the River.* Cambridge, Massachusetts: Harvard University Press. 1994.

Leopold, Luna B. *Water: A Primer.* San Francisco, California: W.H. Freeman and Company. 1974.

Leopold, Luna B., Gordon M. Wolman and John P. Miller. *Fluvial Processes in Geomorphology.* New York, New York: Dover Publications, Inc. 1992.

Littell, John. *First Settlers of Passaic Valley.* Baltimore, Maryland: Genealogical Publishing Company, Inc. 1981.

Ludlum, David M. *The New Jersey Weather Book.* New Brunswick, New Jersey: Rutgers University Press. 1983.

Lurie, Maxine N. and Marc Mappen. *Encyclopedia of New Jersey.* New Brunswick, New Jersey: Rutgers University Press. 1974.

MacNab, John Alleyne. *Song of the Passaic.* New York, New York: Walbridge & Company. 1890.

McCully Betsy. *City at the Water's Edge: A Natural History of New York.* New Brunswick, New Jersey: Rivergate Books, An Imprint of Rutgers University Press. 2007.

Newberry, Joan Duffy. Speech at the International Conference of Victims of Agent Orange. Hanoi, Vietnam. March 2006.

New Jersey Department of Health. *Health Consultation Diamond Alkali Company.* CERCLIS No. NJD980528996. Newark, New Jersey. 1996.

Olsen, Kevin, K. *A Great Conveniency: A Maritime History of the Passaic River, Hackensack River and Newark Bay.* Franklin Tennessee: American History Reprints. 2008.

Rae, John W. *The Mendhams.* Charleston, South Carolina: Arcadia Publishing. 1998.

Richman, Steven M. *The Bridges of New Jersey: Portraits of Garden State Crossings.* New Brunswick, New Jersey: Rutgers University Press. 2005.

Sabini, Meredith, ed. *The Earth Has a Soul: The Nature Writings of C.G. Jung.* Berkeley, California: North Atlantic. 2005.

Salisbury, Rollin D. *The Glacial Geology of New Jersey.* Final Report of the New Jersey State Geologist, Volume V. Trenton, New Jersey: MacCrellish & Quigley, Book and Job Printers. 1902.

Schecter, Arnold and Thomas A. Gasiewicz. eds. *Dioxins and Health.* Hoboken, New Jersey: Wiley Interscience, A John Wiley & Sons, Inc. Publication. 2003.

Schuck, Peter H. *Agent Orange On Trial: Mass Toxic Disasters in the Courts.* Cambridge, Massachusetts: The Belknap Press of Harvard University Press. 1987.

Shaw, Mark. *Melvin Belli: King of the Courtroom.* Fort Lee, New Jersey: Barricade Books, Inc. 2007.

Sheehan, Helen and Richard P. Wedeen, eds. *Toxic Circles: Environmental Hazards From the Workplace Into the Community.* New Brunswick, New Jersey: Rutgers University Press. 1993.

U.S. Army Corps of Engineers, New York District. *Cultural Resources for the Joseph G. Minish Passaic River Waterfront Park and Historic Area, Final Report.* 1998.

U.S. Army Corps of Engineers, New York District. *Flood Protection Feasibility Main Stem Passaic River, Main Report and Environmental Impact Statement.* 1997.

U.S. Army Corps of Engineers, New York District. *Passaic River Flood Damage Reduction Project, Appendix C – Hydrology and Hydraulics.* 1995.

Vermuele, Cornelius Clarkson. *Report on Water-Supply, Water Power, the Flow of Streams and Attendant Phenomena. Volume III of the Final Report of the State Geologist.* Trenton, New Jersey: The John L. Murphy Publishing Company. 1894.

Wacker, Peter O. and Paul G.E. Clemens. *Land Use in Early New Jersey: A Historical Geography.* Newark, New Jersey: The New Jersey Historical Society. 1995.

Wacker, Peter O. *Land & People, A Cultural Geography of Preindustrial New Jersey: Origins and Settlement Patterns.* New Brunswick, New Jersey: Rutgers University Press. 1975.

Wildes, Harry Emerson. *Twin Rivers: The Raritan and the Passaic.* New York, New York: Rhinehart & Company, Inc. 1943.

Williams, William Carlos. *Paterson.* Revised edition prepared by Christopher MacGowan. New York, New York: New Directions. 1992.

Williams, William Carlos. *In the American Grain.* New York, New York: New Directions Publishing Corporation. 1956.

Wolfe, Peter E. *The Geology and Landscapes of New Jersey.* New York, New York. Crane, Russak & Company, Inc. 1977.

World Health Organization. 2010. *Dioxins and Their Effects on Human Health.* Fact sheet No. 225. http://www.who.int/mediacentre/factsheets/fs225/en. November 2011.

Acknowledgments

I RELIED AND IMPOSED UPON QUITE A FEW PEOPLE while working on this book. It seemed that any time I hit a snag in my research or felt my writing energy flag, some family member, friend, colleague or source would inject a welcome dose of enthusiasm or expertise and get me and the book back on track. Each act of support, encouragement and faith helped give birth to *An American River* and I will remain grateful forever. A very special thanks goes to my trusted editor and friend Jean Lenihan, a true mensch; to my agent and fellow Jersey girl Elizabeth Wales; to my soul mate Kate Thompson for book and map design and for the strength and comfort of her love; to my brother John Bruno for his historical sleuthing and beautiful photographs; to Kate (again), John (again), Kathryn Hunt, Mitch Karton and Tom Amorose, my gentle and insightful first readers; Rebecca Farwell, for excerpting the book in *Grist.org;* the University of Washington's Whiteley Center on San Juan Island, where I wrote most of the kayaking passages; Glenda Pearson, who helped me unearth critical Diamond Shamrock court cases; Lynn Rakos for opening up the U.S. Army Corps of Engineers' Passaic River files; Shelley Means, my proofreader; Jim Albrecht, my cosmic consultant; and to my sibs (John, Catherine, Joseph and Paul) for putting me up on countless east coast visits, for sharing their memories of our lives together and for faithfully answering impromptu family trivia questions like "What year did Mom graduate from nursing school?" But the book could never have happened without the invaluable contributions of my primary sources: Andy Willner, Sharon Jaffess, Ella Filippone, Nancy Zak, Arnold Cohen, Carol Johnston, Dr. Robert Chant, Dr. Oliver Hankinson, Michael Gordon, the late Charles Cummings, Otto Zapecza, Dr. Rex Lowe, Len Fariello, Robert Perkins, Kevin Olsen, Nick Sunday, Jerry Willis, Erik King, Carl Alderson, Robert MacFadyen and the late Bob DeVita. It was my privilege and great good fortune to meet them and mine their extraordinary knowledge and generosity, their passion for the environment and their own fascinating life stories. Lastly, thanks again to Carl, my intrepid Passaic River guide, for showing me the other side of the river. I'd follow you anywhere.

Portions of this book first appeared in *New York Woman* magazine, in *A Road of Her Own: Women's Journeys in the West*, and in *Grist.org*.

For news and information about the Passaic River, to get involved with clean up or restoration efforts, or to just get out on the river and have some fun, visit the web sites of these nonprofit organizations, or contact them directly:

The Passaic River Coalition
94 Mt. Bethel Rd
Warren, NJ 07059
908.222.0315
prcwater@aol.com
www.passaicriver.org

NY/NJ Baykeeper Association
52 West Front Street
Keyport, NJ 07735
732.888.9870
mail@nynjbaykeeper.org
www.nynjbaykeeper.org

Nereid Boat Club
350 Riverside Avenue
P.O. Box 1678
Rutherford, NJ 07070
201.438.3995
nereidsec@hotmail.com
www.nereidbc.org

Passaic River Rowing Association
P.O. Box 440
Lyndhurst, NJ 07071
row@prra.org
www.prra.org

Ironbound Community Corporation
Ironbound Administration Office
179 Van Buren St.
Newark, NJ 07105
973.589.3353
info@ironboundcc.org
www.ironboundcc.org

Passaic River Institute
College of Science and Mathematics, Richardson 262
Montclair State University
Montclair, NJ 07043
973.655.7117
pri@montclair.edu
www.csam.montclair.edu/PRI

Passaic Valley Sewerage Commissioners
PVSC Environmental Outreach Program
600 Wilson Avenue
Newark, NJ 07105
973.344.1800
www.pvsc.com

Somerset County Environmental Education Center
190 Lord Stirling Road
Basking Ridge, NJ 07920
908.766.2489
www.somersetcountyparks.org/parksFacilities/eec/EEC.html

Index

Footnotes are indicated by n following the page number. MB refers to Mary Bruno.

Printed in the USA
CPSIA information can be obtained
at www.ICGtesting.com
LVHW011125010823
754044LV00003B/64

9 780615 601793